HANDBOOK OF INSECTS

HANDBOOK OF INSECTS

– Author –
Dr. Praveen Gulati

Wordpen Academics
Publishing Science That Matter

ISBN: 978-81-921738-2-5 (Hardback)

Published by : **WORDPEN ACADEMICS**
4736/23, Ansari Road, Darya Ganj,
New Delhi – 110 002
Phone: +91-011-35004336
E-mail: info@wordpenacademics.com

Printed at **:** **Replika Press Pvt. Ltd.**

Exclusive Distributor: ***Bio-Green Books, New Delhi***

Preface

Most animals on Earth are insects. There are about 800,000 species of animals and about 1 million are insects. That is 80 percent of all animals. Insects have been on Earth for a very long time. Fossils show that insects were here 350 million years ago. Insects have 3 body parts, 2 antennae, and 6 jointed legs. Most adult insects have wings. Flies only have 2 wings. All wings and legs are attached to the thorax body part. All insects are covered by an exoskeleton , a skeleton covering the outside of their bodies like armour. When an insect sheds this covering it is called molting.

Insects also are the most highly developed class of invertebrate animals, with the exception of some mollusks. Insects such as the bees, ants, and termites have elaborate social structures in which the various forms of activity necessary for the feeding, shelter, and reproduction of the colony are divided among individuals especially adapted for the various activities. Also, most insects achieve maturity by metamorphosis rather than by direct growth.

This book contains about 160 insects. The scientific classification, life cycle, habitat, and their control are given in this book. This book may not only provide reference but also serve as a guide and inspiration for future research. The scientists, teachers, entomologists, scholars are expected to find this book indispensable.

Praveen Gulati

Contents

Introduction

Insects (from Latin *insectum*, a calque of Greek *éntomon*, "cut into sections") are a class within the arthropods that have a chitinous exoskeleton, a three-part body (head, thorax, and abdomen), three pairs of jointed legs, compound eyes, and two antennae. They are among the most diverse group of animals on the planet and include more than a million described species and represent more than half of all known living organisms. Insect, common name given to any animal of a class belonging to the arthropod phylum. The insects are thc largest class in the animal world, outnumbering all other animals. At least 800,000 species have been described, and entomologists believe that as many or more remain to be discovered. The class is distributed throughout the world from the polar regions to the tropics and is found on land, in fresh and salt water, and in salt lakes and hot springs. The insects reach their greatest number and variety, however, in the tropics.

In size, the insects also exhibit great variation. Some small parasitic insects are less than 0.025 cm (less than 0.01 in) in length when fully grown, whereas at least one fossilized species related to the modern dragonflies is known to have had a wingspread of more than 60 cm (24 in). The largest insects today are certain stick insects about 30 cm (about 12 in) long and certain moths with wingspans of about 30 cm (about 12 in). Insects also are the most highly developed class of invertebrate animals, with the exception of some mollusks. Insects such as the bees, ants, and termites have elaborate social structures in which the various forms of activity necessary for the feeding, shelter, and reproduction of the colony are divided among-individuals especially adapted for the various activities. Also, most insects achieve maturity by metamorphosis rather than by direct growth. In most species, the individual passes through at least two distinct and dissimilar stages before reaching its adult form. In their living and feeding habits, the insects exhibit extreme variations. Nowhere is this more apparent than in the life cycle of various species. Thus the so-called 17-year locust matures over a period of 13 to 17 years. The ordinary house fly can reach maturity in about ten days, and certain parasitic wasps reach their mature form seven days after the eggs have been laid. In general the insects are very precisely adapted to the environments in which they live, and many species depend on a single variety of plant, usually feeding on one specific portion of the plant such as the leaves, stem, flowers, or roots. The relationship between insect and plant is frequently a necessary one for the growth and reproduction of the plant, as with plants that depend on insects for pollination.

A number of insect species do not feed on living plants but act as scavengers. Some of these species live on decaying vegetable matter and others on dung or the carcasses of animals. The activities of the scavenger insects hasten the decomposition of all kinds of dead organic material. Certain insects also exhibit predation or parasitism, either feeding on other insects or existing on or within the bodies of insect or other animal hosts. Parasitic insects are sometimes parasitic upon parasitic insects, a phenomenon known as hyper-parasitism. In a few instances an insect may be parasitic upon a secondary parasite. A few species of insects, although not strictly parasitic, live at the expense of other insects, with whom they associate closely. An example of this form of relationship is that of th e wax moth, which lives in the hives of bees and feeds on the comb that the bees produce. Sometimes the relation between two species is symbiotic. Thus ant colonies provide food for certain beetles that live with them, and in return the ants consume fluids that have been secreted by the beetles. The number of extant species is estimated at between six and ten million, and potentially represent over 90% of the differing metazoan life forms on Earth. Insects may be found in nearly all environments, although only a small number of species occur in the oceans, a habitat dominated by another arthropod group, the crustaceans. The life cycles of insects vary but most hatch from eggs.

Insect growth is constrained by the inelastic exoskeleton and development involves a series of molts. The immature stages can differ from the adults in structure, habit and habitat and can include a passive pupal stage in those groups that undergo complete metamorphosis. Insects that undergo incomplete metamorphosis lack a pupal stage and adults develop through a series of nymphal stages. The higher level relationship of the hexapoda is unclear. Fossilized insects of enormous size have been found from the Paleozoic Era, including giant dragonflies with wingspans of 55 to 70 cm (22–28 in). The most diverse insect groups appear to have coevolved with flowering plants. Insects typically move about by walking, flying or occasionally swimming. Because it allows for rapid yet stable movement, many insects adopt a tripedal gait in which they walk with their legs touching the ground in alternating triangles. Insects are the only invertebrates to have evolved flight. Many insects spend at least part of their life underwater, with larval adaptations that include gills and some adult insects are aquatic and have adaptations for swimming. Some species, like water striders, are capable of walking on the surface of water. Insects are mostly solitary, but some insects, such as certain bees, ants, and termites are social and live in large, well-organized colonies. Some insects, like earwigs, show maternal care, guarding their eggs and young. Insects can communicate with each other in a variety of ways. Male moths can sense the pheromones of female moths over distances of many kilometres.

Other species communicate with sounds: crickets stridulate, or rub their wings together, to attract a mate and repel other males. Lampyridae in the beetle order Coleoptera communicate with light. Humans regard certain insects as pests and attempt to control them using insecticides and a host of other techniques. Some insects damage crops by feeding on sap, leaves or fruits, a few bite humans and livestock, alive and dead, to feed on blood and some are capable of transmitting diseases to humans, pets and livestock. Many other insects are considered ecologically beneficial and a few provide direct economic benefit. Silkworms and bees have been domesticated by humans for the production of silk and honey, respectively.

Morphology and Physiology

General body Plan

Insects have segmented bodies supported by an exoskeleton, a hard outer covering made mostly of chitin. The segments of the body are organized into three distinctive but interconnected units, or tagmata: a head, a thorax, and an abdomen. The head supports a pair of sensory antennae, a pair of compound eyes, and, if present, one to three simple eyes (or ocelli) and three sets of variously modified appendages that form the mouthparts. The thorax has six segmented legs—one pair each for the prothorax, mesothorax and the metathorax segments making up the thorax—and, if present in the species, two or four wings. The abdomen consists of eleven segments, though in a few species of insects these segments may be fused together or reduced in size. The abdomen also contains most of the digestive, respiratory, excretory and reproductive internal structures. There is considerable variation and many adaptations in the body parts of insects especially wings, legs, antenna, mouth-parts etc.

Exoskeleton

Insect outer skeleton, the cuticle, is made up of two layers: the epicuticle, which is a thin and waxy water resistant outer layer and contains no chitin, and a lower layer called the procuticle. The procuticle is chitinous and much thicker than the epicuticle and has two layers: an outer layer known as the exocuticle and an inner layer known as the endocuticle. The tough and flexible endocuticle is built from numerous layers of fibrous chitin and proteins, criss-crossing each others in a sandwich pattern, while the exocuticle is rigid and hardened.

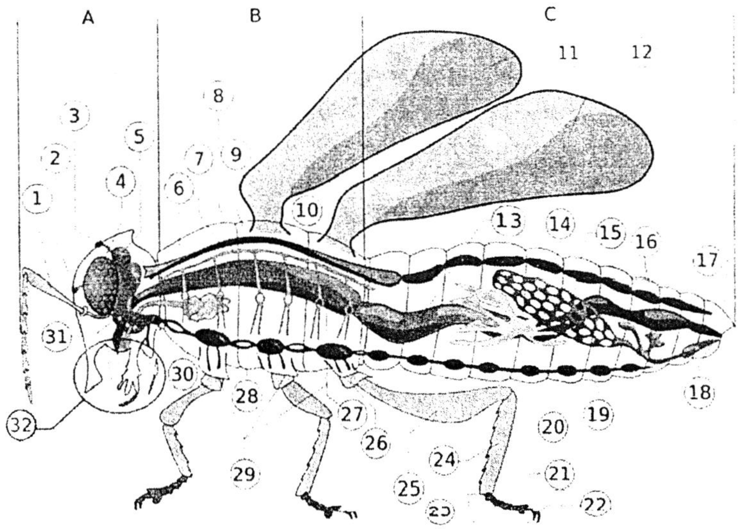

Insect morphology

A- Head B- Thorax C- Abdomen

1. antenna, 2. ocelli (lower), 3. ocelli (upper), 4. compound eye, 5. brain (cerebral ganglia), 6. prothorax, 7. dorsal blood vessel, 8. tracheal tubes (trunk with spiracle), 9. mesothorax, 10. metathorax, 11. forewing, 12. hindwing, 13. mid-gut (stomach), 14. dorsal tube (Heart), 15. ovary, 16. hind-gut (intestine, rectum & anus), 17. anus, 18. oviduct, 19. nerve chord (abdominal ganglia), 20. Malpighian tubes, 21. tarsal pads, 22. claws, 23. tarsus, 24. tibia, 25. femur, 26. trochanter, 27. fore-gut (crop, gizzard), 28. thoracic ganglion, 29. coxa, 30. salivary gland, 31. subesophageal ganglion, 32. mouthparts

The exocuticle is greatly reduced in many soft-bodied insects (e.g., caterpillars), especially during their larval stages. Insects are the only invertebrates to have developed active flight capability, and this has played an important role in their success. These muscles are able to contract multiple times for each single nerve impulse, allowing the wings to beat faster than would ordinarily be possible. Having their muscles attached to their exoskeletons is more efficient and allows more muscle connections; crustaceans also use the same method, though all spiders use hydraulic pressure to extend their legs, a system inherited from their pre-arthropod ancestors. Unlike insects, though, most aquatic crustaceans are biomineralized with calcium carbonate extracted from the water.

Nervous System

The nervous system of an insect can be divided into a brain and a ventral nerve cord. The head capsule is made up of six fused segments, each with a pair of ganglia, or a cluster of nerve cells outside of the brain. The first three pairs of ganglia are fused into the brain, while the three following pairs are fused into a structure of three pairs of ganglia under the insect's esophagus, called the subesophageal ganglion. The thoracic segments have one ganglion on each side, which are connected into a pair, one pair per segment.

This arrangement is also seen in the abdomen but only in the first eight segments. Many species of insects have reduced numbers of ganglia due to fusion or reduction. Some cockroaches have just six ganglia in the abdomen, whereas the wasp *Vespa crabro* has only two in the thorax and three in the abdomen. Some insects, like the house fly *Musca domestica,* have all the body ganglia fused into a single large thoracic ganglion. At least a few insects have nociceptors, cells that detect and transmit sensations of pain.

This was discovered in 2003 by studying the variation in reactions of larvae of the common fruitfly Drosophila to the touch of a heated probe and an unheated one. The larvae reacted to the touch of the heated probe with a stereotypical rolling behavior that was not exhibited when the larvae were touched by the unheated probe. Although nociception has been demonstrated in insects, there is not a consensus that insects feel pain consciously.

Digestive System

An insect uses its digestive system to extract nutrients and other substances from the food it consumes. Most of this food is ingested in the form of macromolecules and other complex substances like proteins, polysaccharides, fats, and nucleic acids. These macromolecules must be broken down by catabolic reactions into smaller molecules like amino acids and simple

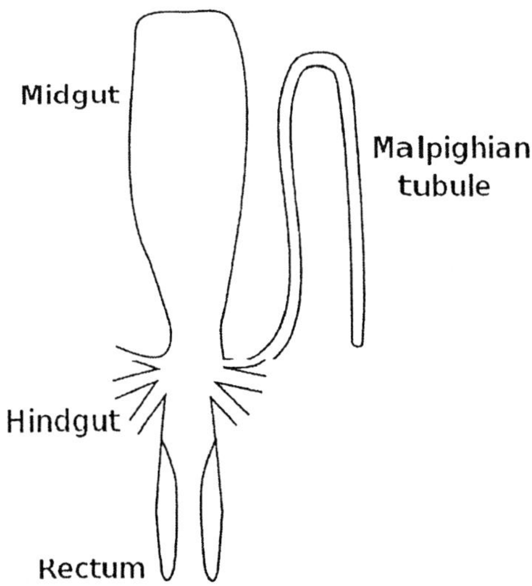

Stylized diagram of insect digestive tract showing malpighian tubule, from an insect of the order Orthoptera

sugars before being used by cells of the body for energy, growth, or reproduction. This break-down process is known as digestion. The main structure of an insect's digestive system is a long enclosed tube called the alimentary canal, which runs lengthwise through the body. The alimentary canal directs food unidirectionally from the mouth to the anus. It has three sections, each of which performs a different process of digestion. In addition to the alimentary canal, insects also have paired salivary glands and salivary reservoirs. These structures usually reside in the thorax, adjacent to the foregut.

The salivary glands in an insect's mouth produce saliva. The salivary ducts lead from the glands to the reservoirs and then forward through the head to an opening called the salivarium, located behind the hypopharynx. By moving its mouthparts the insect can mix its food with saliva. The mixture of saliva and food then travels through the salivary tubes into the mouth, where it begins to break down. Some insects, like flies, have extra-oral digestion. Insects using extra-oral digestion expel digestive enzymes onto their food to break it down. This strategy allows insects to extract a significant proportion of the available nutrients from the food source. The gut is where almost all of insects' digestion takes place. It can be divided into the foregut, midgut and hindgut.

Foregut

Stylized diagram of insect digestive tract showing malpighian tubule, from an insect of the order Orthoptera. The first section of the alimentary canal is the foregut (element 27 in numbered diagram), or stomodaeum. The foregut is line with a cuticular lining made of chitin and proteins as protection from tough food. The foregut includes the buccal cavity (mouth), pharynx, esophagus, and Crop and proventriculus (any part may be highly modified) which both store food and signify when to continue passing onward to the midgut. Here, digestion starts as partially chewed food is broken down by saliva from the salivary glands. As the salivary glands produce fluid and carbohydrate-digesting enzymes (mostly amylases), strong muscles in the pharynx pump fluid into the buccal cavity, lubricating the food like the salivarium does, and helping blood feeders, and xylem and phloem feeders. From there, the pharynx passes food to the esophagus, which could be just a simple tube passing it on to the crop and proventriculus, and then on ward to the midgut, as in most insects. Alternately, the foregut may expand into a very enlarged crop and proventriculus, or the crop could just be a diverticulum, or fluid filled structure, as in some Diptera species.

Midgut

Once food leaves the crop, it passes to the midgut (element 13 in numbered diagram), also known as the mesenteron, where the majority of digestion takes place. Microscopic projections from the midgut wall, called microvilli, increase the surface area of the wall and allow more nutrients to be absorbed; they tend to be close to the origin of the midgut. In some insects, the role of the microvilli and where they are located may vary. For example, specialized microvilli producing digestive enzymes may more likely be near the end of the midgut, and absorption near the origin or beginning of the midgut.

Hindgut

In the hindgut (element 16 in numbered diagram), or proctodaeum, undigested food particles are joined by uric acid to form fecal pellets. The rectum absorbs 90% of the water in these fecal pellets, and the dry pellet is then eliminated through the anus (element 17), completing the process of digestion. The uric acid is formed using hemolymph waste products diffused from the Malpighian tubules (element 20). It is then emptied directly into the alimentary canal, at the junction between the midgut and hindgut. The number of Malpighian tubules possessed by a given insect varies between species, ranging from only two tubules in some insects to over 100 tubules in others.

Social Insects

Characteristically an insect society is formed of a parent or parents and a large number of offspring. The individual members of the society are divided into groups, each having a specialized function and often exhibiting markedly different bodily structures. For discussion of the organization of typical insect societies. One of the most interesting forms of insect behavior is exhibited by the social insects, which, unlike the majority of insect species, live in organized groups. The social insects include about 800 species of wasps, 500 species of bees, and the ants and termites.

Insect Anatomy

All insects have three pairs of legs, each pair growing from a different part of the thorax, called, from front to back, the prothorax, the mesothorax, and the metathorax. Many larvae have, in addition, several pairs of leglike appendages called struts, or prolegs. The forms of the legs vary, depending on their uses, but all insect legs are made up of five parts. In winged insects, the wings, usually four in number, grow from the thorax between the mesothorax and the metathorax. The upper and lower membranes of the wings cover a network of sclerotized tubes, called veins, that stiffen the wing. The pattern of veins of the wings is characteristic of most insect species and is extensively used by entomologists as a basis for classification. Although the superficial appearance of insects is extremely varied, certain characteristics of their anatomy are common to the entire class. All mature insects have bodies composed of three parts: head, thorax, and abdomen (the abdomen and thorax are not always differentiated in larvae).

Each of these parts is composed of a number of segments. The head is made up of several segments, usually so fused they are scarcely differentiated. On the head are two antennae; a pair of jaws, or mandibles; a pair of auxiliary jaws, or maxillae, that in turn bears a pair of palps; and a fused second pair of accessory jaws, the labium, that also bears a pair of palps. The antennae, usually attached to the anterior part of the head, are segmented. In some insects, the antennae carry organs of smell as well as organs of touch. The mandibles are large, heavy jaws on either side of the mouth. They close horizontally and are used for grasping food and crushing it. The maxillae, or inner jaws, are lighter in structure. The mouths of many insects are adapted for piercing and sucking rather than for biting. The eyes of the insect are also situated on the head. Insect abdomens usually have 10 or 11 clearly defined segments. In all cases the anal opening is located on the last segment; in some species, such as the mayflies, a pair of feelers, called cerci, is also present on this segment.

The abdomen is devoid of legs. In female insects, it contains the egg-laying organ, or ovipositor, which may be modified into a sting, saw, or drill for depositing the eggs in the bodies of plants or animals. Insect sexual organs arise from the eighth and ninth segments of the abdomen. Insects have an external rather than an internal skeleton; this exoskeleton is a rough integument formed by the hardening of the outer layer of the body through impregnation with pigments and polymerization of proteins, a process known as sclerotization. The exoskeleton at the joints does not become sclerotized and therefore remains flexible.

Flight

Insects are the only group of invertebrates to have developed flight. The evolution of insect wings has been a subject of debate. Some entomologists suggest that the wings are from paranotal lobes, or extensions from the insect's exoskeleton called the nota, called the *paranotal theory*. Other theories are based on a pleural origin. The pleuron is membrane on the sides of the thorax. These theories include suggestions that wings originated from modified gills, spiracular flaps or as from an appendage of the epicoxa. The *epicoxal theory* suggests the insect wings are modified epicoxal exites, a modified appendage at the base of the legs or coxa. In the Carboniferous age, some of the *Meganeura* dragonflies had as much as a 50 cm (20 in) wide wingspan. The appearance of gigantic insects has been found to be consistent with high atmospheric oxygen. The respiratory system of insects constrains their size, however the high oxygen in the atmosphere allowed larger sizes. The largest flying insects today are much smaller and include several moth species such as the Atlas moth and the White Witch (*Thysania agrippina*). Insect flight has been a topic of great interest in aerodynamics due partly to the inability of steady-state theories to explain the lift generated by the tiny wings of insects.

Unlike birds, many small insects are swept along by the prevailing winds although many of the larger insects are known to make migrations. Aphids are known to be transported long distances by low-level jet streams. Most insects possess wings during at least part of their life cycles. Insect wings are large folds in the exoskeleton composed of two sheets of cuticle permeated with stiff supportive veins. The wings are powered by two sets of muscles that independently drive the upstroke and downstroke of the wing movement. The frequency of wingbeats ranges from four beats per second in butterflies to nearly 1000 beats per second in some gnats. Wings not only move up and down but they also move forward and backward in an ellipse or figure eight pattern that provides both lift and thrust. Given their shape, speed, and stroke pattern, it has never been clearly understood how insect wings can generate enough lift to sustain flight. Recently scientists discovered that insects generate a vortex, or spiral air motion, along the leading edge of their wings. This vortex flows out toward the wing tip in widening spirals. The whirling cylinder of air above the insect provides the extra lift that makes flight possible.

Respiration

Certain species of insects breathe through the body wall, by diffusion, but in general the respiratory system of members of this class consists of a network of tubes, or tracheae, that carry air throughout the body to smaller tubelets or tracheoles with which all the organs of the body are supplied. In the tracheoles the oxygen from the air diffuses into the bloodstream, and carbon dioxide from the blood diffuses into the air. The exterior openings of the tracheae are called spiracles. The spiracles are situated on the sides of the insect and are usually 20 in number (10 pairs), four on the thorax and 16 on the abdomen. Some water-breathing insects have gill-like structures.

Circulation

The circulatory system of insects is simple. The entire body cavity is filled with blood that is kept in circulation by means of a simple heart. This heart is a tube, open at both ends, that runs the entire length of the body under the exoskeleton along the back of the insect. The walls of the heart can contract to force the blood forward through the heart and out into the body cavity.

Reproduction and Development

The majority of insects hatch from eggs. The fertilization and development takes place inside the egg, enclosed by a shell (chorion). Some species of insects, like the cockroach *Blaptica dubia,* as well as juvenile aphids and tsetse flies, are ovoviviparous. The eggs of ovoviviparous animals develop entirely inside the female, and then hatch immediately upon being laid. Some other species, such as those in the genus of cockroaches known as *Diploptera,* are viviparous, and thus gestate inside the mother and are born alive. Some insects, like parasitic wasps, show polyembryony, where a single fertilized egg divides into many and in some cases thousands of separate embryos. Other developmental and reproductive variations include haplodiploidy, polymorphism, paedomor-phosis or peramorphosis, sexual dimorphism, parthenogenesis and more rarely hermaphroditism. In haplodiploidy, which is a type of sex-determination system, the offspring's sex is determined by the number of sets of chromosomes an individual receives. This system is typical in bees and wasps. Polymorphism is the where a species may have different *morphs* or *forms,* as in the oblong winged katydid, which has four different varieties: green, pink, and yellow or tan. Some insects may retain phenotypes that are normally only seen in juveniles; this is called paedomorphosis. In peramorphosis, an opposite sort of phenomenon, insects take on previously unseen traits after they have matured into adults.

Many insects display sexual dimorphism, in which males and females have notably different appearances, such as the moth *Orgyia recens* as an exemplar of sexual dimorphism in insects. Some insects use parthenogenesis, a process in which the female can reproduce and give birth without having the eggs fertilized by a male. Many aphids undergo a form of parthenogenesis, called cyclical parthenogenesis, in which they alternate between one or many generations of asexual and sexual reproduction. In summer, aphids are generally female and parthenogenetic; in the autumn, males may be produced for sexual reproduction. Other insects produced by parthenogenesis are bees, wasps, and ants, in which they spawn males. However, overall, most individuals are female, which are produced by fertilization. The males are haploid and the females are diploid. More rarely, some insects display hermaphroditism, in which a given individual has both male and female reproductive organs. Insect life-histories show adaptations to withstand cold and dry conditions. Some temperate region insects are capable of activity during winter, while some others migrate to a warmer climate or go into a state of torpor. Still other insects have evolved mechanisms of diapause that allow eggs or pupae to survive these conditions

Metamorphosis

Metamorphosis in insects is the biological process of development all insects must undergo. There are two forms of metamorphosis: incomplete metamorphosis and complete metamorphosis. One of the characteristics of the development of insects from birth to maturity is metamorphosis, the change through one or more distinctive immature body forms to the imago, or adult body form. Metamorphosis of some kind occurs in most insects, although in a few species, such as the bristletails, the newborn insect is essentially similar in form to the imago.

Incomplete Metamorphosis

Insects that show hemimetabolism, or incomplete metamorphosis, change gradually by undergoing a series of molts. An insect molts when it outgrows its exoskeleton, which does not stretch and would otherwise restrict the insect's growth. The molting process begins as the insect's epidermis secretes a new epicuticle. After this new epicuticle is secreted, the epidermis releases a mixture of enzymes that digests the endocuticle and thus detaches the old cuticle. When this stage is complete, the insect makes its body swell by taking in a large quantity of water or air, which makes the old cuticle split along predefined weaknesses where the old exocuticle was thinnest. Other arthropods have a much different process and only molt; though must accommodate for the difference in exoskeleton structure and make up with other enzymes.

Immature insects that go through incomplete metamorphosis are called nymphs or in the case of dragonflies and damselflies as naiads. Nymphs are similar in form to the adult except for the presence of wings, which are not developed until adulthood. With each molt, nymphs grow larger and become more similar in appearance to adult insects.

Complete Metamorphosis

Holometabolism, or complete metamorphosis, is where the insect changes all in four stages, an egg or embryo, a larva, a pupa, and the adult or imago. In these species, egg hatches to produce a larva, which is generally worm-like in form. This worm-like form can be one of several varieties: eruciform (caterpillar-like), scarabaeiform (grub-like), campodeiform (elongated, flattened, and active), elateriform (wireworm-like) or vermiform (maggot-like). The larva grows and eventually becomes a pupa, a stage marked by reduced movement and often sealed within a cocoon. There are three types of pupae: obtect, exarate or coarctate. Obtect pupae are compact, with the legs and other appendages enclosed. Exarate pupae have their legs and other appendages free and extended. Coarctate pupae develop inside the larval skin. Insects undergo considerable change in form during the pupal stage, and emerge as adults. Butterflies are a well known example of an insects that undergo complete metamorphosis, although most insects use this life cycle. Some insects have evolved this system to hypermetamorphosis.

Some of the oldest and most successful insect groups, such Endopterygota, use a system of complete metamorphosis. Strangely though, complete metamorphosis is unique to certain insect orders, like Diptera, Lepidoptera, and Hymenoptera, and no other arthropods undergo it, but incomplete metamorphosis.

Senses and Communication

Many insects possess very sensitive and/or specialized organs of perception. Some insects such as bees can perceive ultraviolet wavelengths, or detect polarized light, while the antennae of male moths can detect the pheromones of female moths over distances of many kilometers. There is a pronounced tendency for there to be a trade-off between visual acuity and chemical or tactile acuity, such that most insects with well-developed eyes have reduced or simple antennae, and vice-versa. There are a variety of different mechanisms by which insects perceive sound, while the patterns are not universal, insects can generally hear sound if they can produce it.

Different insect species can have varying hearing, though most insects can hear only a narrow range of frequencies related to the frequency of the sounds they can produce. Mosquitoes have been found to hear up to 2 MHz, and some grasshoppers can hear up to 50 MHz. Certain predatory and parasitic insects can detect the characteristic sounds made by their prey or hosts, respectively. For instance, some nocturnal moths can perceive the ultrasonic emissions of bats, which helps them avoid predation. Insects that feed on blood have special sensory structures that can detect infrared emissions, and use them to home in on their hosts. Some insects display a rudimentary sense of numbers, such as the solitary wasps that prey upon a single species. The mother wasp lays her eggs in individual cells and provides each egg with a number of live caterpillars on which the young feed when hatched.

Some species of wasp always provide five, others twelve, and others as high as twenty-four caterpillars per cell. The number of caterpillars is different among species, but always the same for each sex of larva. The male solitary wasp in the genus *Eumenes* is smaller than the female, so the mother of one species supplies him with only five caterpillars; the larger female receives ten caterpillars in her cell.

Light Production and Vision

A few insects, such as members of the families Poduridae and Onychiuridae (Collembola), Mycetophilidae (Diptera), and the beetle families Lampyridae, Phengodidae, Elateridae and

Staphylinidae are bioluminescent. The most familiar group are the fireflies, beetles of the family Lampyridae. Some species are able to control this light generation to produce flashes. The function varies with some species using them to attract mates, while others use them to lure prey. Cave dwelling larvae of *Arachnocampa* (Mycetophilidae, Fungus gnats) glow to lure small flying insects into sticky strands of silk. Some fireflies of the genus *Photuris* mimic the flashing of female *Photinus* species to attract males of that species, which are then captured and devoured.

The colours of emitted light vary from dull blue (*Orfelia fultoni*, Mycetophilidae) to the familiar greens and the rare reds (*Phrixothrix tiemanni*, Phengodidae). Most insects, except some species of cave dwelling crickets, are able to perceive light and dark. Many species have acute vision capable of detecting minute movements. The eyes include simple eyes or ocelli as well as compound eyes of varying sizes. Many species are able to detect light in the infrared, ultraviolet and the visible light wavelengths. Colour vision has been demonstrated in many species and phylogenetic analysis suggests that UV-green-blue trichromacy existed from at least the Devonian period between 416 and 359 million years ago.

Sound Production and Hearing

Insects were the earliest organisms to produce and sense sounds. Insects make sounds mostly by mechanical action of appendages. In grasshoppers and crickets, this is achieved by stridulation. Cicadas make the loudest sounds among the insects by producing and amplifying sounds with special modifications to their body and musculature. The African cicada *Brevisana brevis* has been measured at 106.7 decibels at a distance of 50 cm (20 in). Some insects, such as the hawk moths and Hedylid butterflies, can hear ultrasound and take evasive action when they sense that they have been detected by bats. Some moths produce ultrasonic clicks that were once thought to have a role in jamming bat echolocation. The ultrasonic clicks were subsequently found to be produced mostly by unpalatable moths to warn bats, just as warning colorations are used against predators that hunt by sight.Some otherwise palatable moths have evolved to mimic these calls.

Very low sounds are also produced in various species of Coleoptera, Hymenoptera, Lepidoptera, Mantodea, and Neuroptera. These low sounds are simply the sounds made by the insect's movement. Through microscopic stridulatory structures located on the insect's muscles and joints, the normal sounds of the insect moving are amplified and can be used to warn or communicate with other insects. Most sound-making insects also have tympanal organs that can perceive airborne sounds. Some species in Hemiptera, such as the corixids (water boatmen), are known to communicate via underwater sounds.

Most insects are also able to sense vibrations transmitted through surfaces. For example, an insect is caught in a spider web and struggles to escape. The vibrations it produces are sensed by the spider, who is alerted to its presence. Through these vibrations, the spider can tell where on the web the insect is located, as well as how big it is. Communication using surface-borne vibrational signals is more widespread among insects because of size constraints in producing air-borne sounds. Insects cannot effectively produce low-frequency sounds, and high-frequency sounds tend to disperse more in a dense environment (such as foliage), so insects living in such environments communicate primarily using substrate-borne vibrations.

The mechanisms of production of vibrational signals are just as diverse as those for producing sound in insects. Some species use vibrations for communicating within members of the same species, such as to attract mates as in the songs of the shield bug *Nezara viridula*. Vibrations can also be used to communicate between entirely different species; lycaenid (gossamer-winged butterfly) caterpillars which are myrmecophilous (living in a mutualistic association with ants) communicate with ants in this way. The Madagascar hissing cockroach has the ability to press air through its spiracles to make a hissing noise as a sign of aggression; the Death's-head Hawkmoth makes a squeaking noise by forcing air out of their pharynx when agitated, which may also reduce aggressive worker honey bee behavior when the two are in close proximity.

Chemical Communication

In addition to the use of sound for communication, a wide range of insects have evolved chemical means for communication. These chemicals, termed semiochemicals, are often derived from plant metabolites include those meant to attract, repel and provide other kinds of information. Pheromones, a type of semiochemical, are used for attracting mates of the opposite sex, for aggregating conspecific individuals of both sexes, for deterring other individuals from approaching, to mark a trail, and to trigger aggression in nearby individuals. Allomonea benefit their producer by the effect they have upon the receiver. Kairomones benefit their receiver instead of their producer.

Synomones benefit the producer and the receiver. While some chemicals are targeted at individuals of the same species, others are used for communication across species. The use of scents is especially well known to have developed in social insects.

Fossil Insects

Insects were the first animals on Earth to develop flight, which is probably one reason they've been so successful. In sheer numbers, insects dominate the animal kingdom, with over 750,000 living species described. This is more than all the other living animal species combined, and scientists estimate that there are perhaps at least a million more insect species that have not yet been described. Insects were probably similarly abundant in the past, but the fossil record of insects is limited, with only about 7,000 species known.

The scarcity of insect fossils (compared to their present, and probable past, diversity) is a consequence of their relative fragility, lack of hard parts, and the fact that they lived on land.

Rapid burial is essential for preservation, and this is more likely to happen in and around water than on land.

Scientific Classification

The Apterygota are divided into four orders: the Protura, a group of very tiny, blind insects; the Thysanura, which includes the silverfish; the Diplura, a small group that contains the largest (about 5 cm/2 in long) of the Apterygota, an insect of the genus Heterojapyx; and the Collembola, which includes the springtail. The class of insects is arranged in various ways by different entomologists. The whole class Insecta is divided into two subclasses: the Apterygota, wingless insects; and the Pterygota, including the great bulk of insects, most of which have wings in the imago form.

The Pterygota are divided into 27 orders: the Ephemeroptera, which includes the mayflies; the Plecoptera, the stone flies; the Odonata, the dragonflies and damselflies; the Grylloblattodea, a small wingless order; the Orthoptera, the katydids, crickets, and grasshoppers (locusts); the Phasmida, the walkingsticks; the Thysanoptera, the thrips; the Dermaptera, the earwigs; the Mantodea, the mantids; the Blattaria, the roaches (sometimes placed together with the mantids in the order Dictyoptera); the Isoptera, the termites; the Embioptera, a small group of subsocial insects living in tropical and subtropical regions; the Psocoptera, the bark lice and book lice; the Phthiraptera, the bird lice and true lice (sometimes separated into the two orders Mallophaga and Siphonculata); the Zoraptera, of which only one termitelike genus (including about 20 species) is known; the Megaloptera, the alderflies and dobsonflies; the Raphidiodea, the snake flies; the Neuroptera, the antlions and lacewings; the Mecoptera, the scorpion flies; the Trichoptera, the caddisflies; the Lepidoptera, the butterflies and moths; the Diptera, the mosquitoes, gnats, and true flies; the Siphonaptera, the fleas; the Coleoptera, the beetles; the Strepsiptera, a group of tiny insects parasitic on other insects; the Hymenoptera, including the ants, bees, wasps, hornets, and ichneumons, and chalcids; and the Hemiptera.

Ant

Kingdom: Animalia, **Phylum**: Arthropoda, **Class**: Insecta **Order**: Hymenoptera, **Suborder**: Apocrita, **Superfamily**: Vespoidea, **Family**: **Formicidae** Latreille, 1809.

Ants are social insects of the family **Formicidae** and, along with the related wasps and bees, belong to the order Hymenoptera. Ants evolved from wasp-like ancestors in the mid-Cretaceous period between 110 and 130 million years ago and diversified after the rise of flowering plants. More than 12,500 out of an estimated total of 22,000 species have been classified. They are easily identified by their elbowed antennae and a distinctive node-like structure that forms a slender waist.

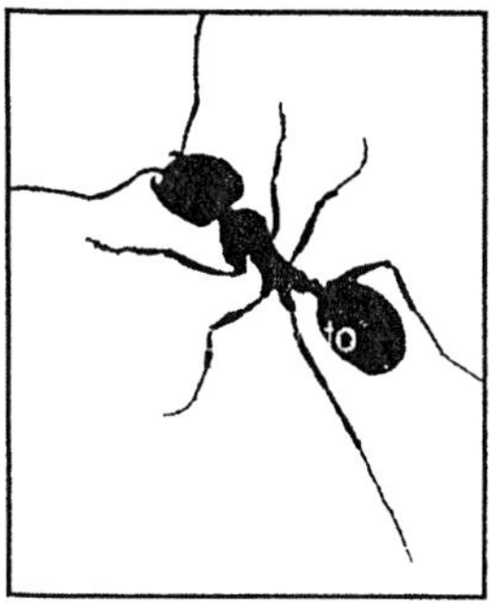

Although ants are frustrating when they get into your home or when you're having a picnic, ants do help the environment. They are social insects, which means they live in large colonies or groups. Depending on the species, ant colonies can consist of millions of ants. There are three kinds of ants in a colony: The queen, the female workers, and males. The queen and the males have wings, while the workers don't have wings. The queen is the only ant that can lay eggs.

The male ant's job is to mate with future queen ants and they do not live very long afterwards. Once the queen grows to adulthood, she spends the rest of her life laying eggs! Depending on the species, a colony may have one queen or many queens.

Ant colonies also have soldier ants that protect the queen, defend the colony, gather or kill food, and attack enemy colonies in search for food and nesting space. If they defeat another ant colony, they take away eggs of the defeated ant colony. When the eggs hatch, the new ants become the "slave" ants for the colony. Some jobs of the colony include taking care of the eggs and babies, gathering food for the colony and building the anthills or mounds. Ants have colonised almost every landmass on Earth. The only places lacking indigenous ants are Antarctica and a few remote or inhospitable islands. Ants thrive in most ecosystems, and may form 15–25% of the terrestrial animal biomass. Their success in so many environments has been attributed to their social organisation and their ability to modify habitats, tap resources, and defend themselves. Their long co-evolution with other species has led to mimetic, commensal, parasitic, and mutualistic relationships.

Ant societies have division of labour, communication between individuals, and an ability to solve complex problems. These parallels with human societies have long been an inspiration and subject of study. Many human cultures make use of ants in cuisine, medication and rituals. Some species are valued in their role as biological pest control agents. However, their ability to exploit resources brings ants into conflict with humans, as they can damage crops and invade buildings. Some species, such as the red imported fire ant, are regarded as invasive species, establishing themselves in areas where they are accidentally introduced.

Characteristics

Ants are closely related to wasps, as can be seen in their similar body structures-the abdomen is joined to the thorax by a slender stalk, or pedicel. The pedicel may be enlarged into one or two knobs. The antennae are typically elbowed or jointed in the middle. Some species of ants possess a functional sting that the workers use to defend the colony or themselves. Many species secrete formic acid, a potent repellent. The jaws of worker ants are used for many tasks, including defense, nest building, and larval care. The form of the jaws is often very specific to the type of work done by the particular worker.

Habitat

While some ants form complex nests and galleries, other species are nomadic and do not build permanent structures. Various species may form subterranean nests or build them on trees. Nests can be found in the ground with craters or mounds around the entrance, under stones or logs, in logs, hollow stems, even acorns. The materials used for construction include soil and plant matter, and they are highly selective of the nest site; Temnothorax albipennis will avoid sites with dead ants as these may be indicators of pests or disease. They are also quick to abandon established nest sites at the first sign of these threats. Some of the more advanced ants are the army ants and driver ants, from South America and Africa respectively. Unlike most species which have permanent nests, army and driver ants do not form permanent nests, but instead alternate between nomadic stages and stages where the workers form a temporary nest (bivouac) out of their own bodies. Colonies reproduce either through nuptial flights as described above, or by fission, where a group of workers simply dig a new hole and raise new queens. Colony members are distinguished by smell, and other intruders are usually attacked. Weaver ants (Oecophylla) build nests in trees by attaching leaves together, first pulling them together with bridges of workers and then sewing them together by pressing silk-producing larvae against them in alternation.

Development and Reproduction

The life of an ant starts from an egg. If the egg is fertilised, the progeny will be female (diploid); if not, it will be male (haploid). Ants develop by complete metamorphosis with the larval stages passing through a pupal stage before emerging as an adult. The larva is largely immobile and is fed and cared for by workers. Food is given to the larvae by trophallaxis, a process in which an ant regurgitates liquid food held in its crop. This is also how adults share food, stored in the "social stomach", among themselves. Larvae may also be provided with solid food such as trophic eggs, pieces of prey and seeds brought back by foraging workers and may even be transported directly to captured prey in some species. The larvae grow through a series of moults and enter the pupal stage. The pupa has the appendages free and not fused to the body as in a butterfly pupa. The differentiation into queens and workers (which are both female), and different castes of workers (when they exist), is influenced in some species by the nutrition the larvae obtain.

Genetic influences and the control of gene expression by the developmental environment are complex and the determination of caste continues to be a subject of research. The developmental environment Larvae and pupae need to be kept at fairly constant temperatures to ensure proper development, and so are often moved around the various brood chambers within the colony.

A new worker spends the first few days of its adult life caring for the queen and young. It then graduates to digging and other nest work, and later to defending the nest and foraging. These changes are sometimes fairly sudden, and define what are called temporal castes. An explanation for the sequence is suggested by the high casualties involved in foraging, making it an acceptable risk only for ants that are older and are likely to die soon of natural causes.

Behaviour and Ecology

Communication

Weaver ants collaborating to dismember a red ant (the two at the extremities are pulling the red ant, while the middle one cuts the red ant until it snaps). Ants communicate with each other using pheromones.

These chemical signals are more developed in ants than in other hymenopteran groups. Like other insects, ants perceive smells with their long, thin and mobile antennae. The paired antennae provide information about the direction and intensity of scents. Since most ants live on the ground, they use the soil surface to leave pheromone trails that can be followed by other ants. In species that forage in groups, a forager that finds food marks a trail on the way back to the colony; this trail is followed by other ants, these ants then reinforce the trail when they head back with food to the colony. When the food source is exhausted, no new trails are marked by returning ants and the scent slowly dissipates. This behaviour helps ants deal with changes in their environment. For instance, when an established path to a food source is blocked by an obstacle, the foragers leave the path to explore new routes. If an ant is successful, it leaves a new trail marking the shortest route on its return. Successful trails are followed by more ants, reinforcing better routes and gradually finding the best path. Ants use pheromones for more than just making trails. A crushed ant emits an alarm pheromone that sends nearby ants into an attack frenzy and attracts more ants from further away. Several ant species even use "propaganda pheromones" to confuse enemy ants and make them fight among themselves.

Pheromones are produced by a wide range of structures including Dufour's glands, poison glands and glands on the hindgut, pygidium, rectum, sternum and hind tibia. Pheromones are also exchanged mixed with food and passed by trophallaxis, transferring information within the colony. This allows other ants to detect what task group (*e.g.*, foraging or nest maintenance) other colony members belong to. In ant species with queen castes, workers begin to raise new queens in the colony when the dominant queen stops producing a specific pheromone. Some ants produce sounds by stridulation, using the gaster segments and their mandibles. Sounds may be used to communicate with colony members or with other species.

Ant Pest

Some ant species are considered pests, and because of the adaptive nature of ant colonies, eliminating the entire colony is nearly impossible. Pest management is therefore a matter of controlling local populations, instead of eliminating an entire colony, and most attempts at control are temporary solutions. Ants classified as pests include the pavement ant, yellow crazy ant, sugar ants, the Pharaoh ant, carpenter ants, Argentine ant, odorous house ants, red imported fire ant and European fire ant. Populations are controlled using insecticide baits, either in granule or liquid formulations. Bait is gathered by the ants as food and brought back to the nest where the poison is inadvertently spread to other colony members through trophallaxis. Boric acid and borax are often used as insecticides that are relatively safe for humans. Bait may be broadcast over a large area to control species like the red fire ant that occupy large areas. Nests of red fire ants may be destroyed by following the ants' trails back to the nest and then pouring boiling water into it to kill the queen. In the United States, one of the most harmful pests among the ants is the red imported fire ant, a small stinging species that was introduced accidentally from South America. The large mound communities of this pest disrupt mechanized farming, and damage crops, and raiding workers injure or kill livestock and wildlife. The little black ant and its close relative, the Pharaoh ant, infest households in the United States. In the northern states, the carpenter ant is another common intruder in and around dwellings. Colonies of this structural pest usually gnaw out the galleries and chambers of their nests in wood initially infested by beetles. Also common in the north is the pavement ant, native to some Middle East countries but now found worldwide. Ants can be a nuisance simply because of their stings and bites and sometimes pugnacious behavior. Some species sting and bite simultaneously, rubbing caustic formic acid into the painful wounds they inflict. Wearing proper clothing can help to avoid ant bites and stings. Ants may pose a health threat to individuals that are allergic to insect stings.

The red imported fire ant is the most notorious North American species in this regard. Harvester ants are responsible for many ant attacks. In Australia, bulldog ant workers can be up to 2.5 cm (1 in) in length, and they have a formidable sting.

Similarly impressive in size is the bullet ant of tropical South and Central America, which reaches 2.2 cm (0.87 in) in length. Burchell's army ant is a widely known tropical species. Army ants inhabit tropical areas. The workers include the major, or soldier, type, which has huge hook-shaped mandibles and a potent sting. Each day, colonies of this species organize a swarm raid that may have a front almost 14 m (more than 45 ft) wide and that advances at a rate of about 30 cm (about 12 in) a minute, although individual ants move much faster. These ants capture, tear, and carry back to their temporary nest any prey that cannot escape them. Their victims are usually other insects or spiders but at times, nestling birds, cornered snakes, or other small vertebrates are killed by stinging. Driver ants in one African genus carry out comparable swarm raids, which usually begin in the early evening. Small forest mammals or domestic stock cornered in pens and unable to escape the ants are sometimes included in the kill.

The economic significance of ants as pests is difficult to judge. Carpenter ants, which are considered destructive to wood, actually may contribute to forest economy by hastening the breakdown and recycling of timber previously infested by other insects. Also, seed-gathering ants may be destructive to agriculture when they become excessively numerous around grain fields or storage centers, but they also may suppress beetles and other grain pests. Ants such as aphid-tending species are frequent pests around lawns and gardens; however, the great benefit of these and other ants in aerating and mixing the soil also must be considered. Driver and army ants are efficient exterminators of other, more damaging insects and in some regions are temporarily allowed to enter human dwellings to clean up such pests. Most of the worst pest ants are ecologically displaced species that thrive in disturbed or artificial habitats created by humans.

The African driver ants that carry out swarm raids are classified in the genus Dorylus. Bulldog ants are in the genus Myrmecia, and the bullet ant is *Paraponera clavata.* Ants make up the family Formicidae of the order Hymenoptera. The red harvester ant is classified as *Pogonomyrmex barbatus.* The ants of the southeastern United States and tropical America that cultivate a fungus in their nests belong to the genus Atta. Honeypot ants are classified in the genus *Myrmecocystus.* The Pharaoh ant is classified as Monomorium pharaonis, the Amazon ant as Polyergus breviceps, the red imported fire ant as *Solenopsis invicta,* and the little black ant as *Monomorium minimum.* The carpenter ant of the northern United States is Campanotus pennsyl-vannicus, the pavement ant is *Tetramorium caespitum,* and the most widely known of the American army ants is *Eciton burchelli.*

ARGENTINE ANT

Common Name: Argentine Ant, **Kingdom**: Animalia, **Phylum**: Arthropoda, **Class**: Insecta, **Order**: Hymenoptera, **Family**: Formicidae, **Species**: *Linepithema humile*

Size: 1/16"to 1/4", **Shape**: Segmented, Oval, **Color**: Dark brown to black and shiny, **Legs**: 6, **Wings**: Varies, **Antenna**: Yes

This species of ant is native to Argentina and Brazil and was probably introduced to the United States in freight ships around the 1890's. These ants can be found in southern states and in California, Illinois, Maryland, Missouri, Oregon and Washington.

Diet

Argentine Ants prefer sweet substances but will eat almost anything including meats, eggs, oils and fats. Also, when foraging for food, Argentine ants leave pheromone trails everywhere they go, instead of just from nest to food source. This habit ensures they do not waste time visiting the same area twice. While in other ant species worker ants are primarily responsible for gather food, Argentine queens also assist with foraging for food.

Habitat

Argentine ant colonies are located in wet environments near a food source. These colonies can grow to monumental size, sometimes covering entire habitats, such as an entire garden or your whole back yard.

Impact

Argentine ants do not pose a health threat, but they can contaminate food by leaving their bodily waste behind.

Prevention

- Eliminate standing water. Pests, such as ants, mosquitoes and termites, are attracted to moisture.
- Keep tree branches and other plants cut back from the house. Sometimes pests use these branches to get into your home.
- Make sure that there are no cracks or little openings around the bottom of your house. Sometimes pests use these to get into your home.
- Make sure that firewood and building materials are not stored next to your home. Pests like to build nests in stacks of wood.

CARPENTER ANTS

Common Name: Carpenter Ant, **Kingdom**: Animalia, **Phylum**: Arthropoda, **Class**: Insecta, **Order**: Hymenoptera, **Family**: Formicidae, **Species**: *Camponotus*.

Size: 5/8", **Shape**: Oval, **Color**: Range in color from red to black, **Legs**: 6, **Wings**: Varies, **Antenna**: Yes.

Carpenter Ants get their name because they build their nests in wood. That can cause significant damage to the wood in your house. There are nine types of carpenter ants throughout the U.S. and range in size from one-quarter inch (about the width of a pencil) for a worker carpenter ant to three-quarters of an inch (about the size of a quarter) for a queen carpenter ant.Each colony is established by a single, fertilized queen. She starts her nest in a cavity in wood, where she raises her first brood of workers. She feeds them saliva and does not leave the nest or feed herself during this time.When they are ready, those workers then get the job of gathering food to feed the next generation. Once mature, this first generation of worker ants work to increase the food supply for the colony. The colony population grows very rapidly. A colony can eventually produce 2,000 or more workers.

Diet

Carpenter ants do not eat the wood they remove during their nest-building activities, but deposit it outside entrances to the colony in small piles. The diet of carpenter ants includes living and dead insects, meat, fats and sugary foods of all kinds, including honeydew and nectar from plants.

Habitat

Carpenter ants build nests anywhere they can find water and moldy or damp wood, such as tree stumps, firewood or in the plants around your house. Carpenter ants also build nests inside, usually entering buildings through wet, damaged wood, although it isn't uncommon for them to adapt to drier environments.

Impact

Carpenter ants don't carry disease, but when building a nest inside a home, Carpenter Ants dig smooth tunnels inside the wood. These tunnels weaken the wood and potentially damage the wood that keeps the house standing. This kind of damage can be very expensive to fix.

Prevention

- Eliminate standing water. Pests, such as ants, mosquitoes and termites, are attracted to moisture.
- Keep tree branches and other plants cut back from the house. Sometimes pests use these branches to get into your home.
- Make sure that there are no cracks or little openings around the bottom of your house. Sometimes pests use these to get into your home.
- Make sure that firewood and building materials are not stored next to your home. Pests like to build nests in stacks of wood.

ODOROUS HOUSE ANTS

Common Name: Odorous House Ant, **Kingdom**: Animalia, **Phylum**: Arthropoda, **Class**: Insecta, **Order**: Hymenoptera, **Family**: Formicidae, **Species**: *Tapinoma sessile*

Size: 1/16"to 1/8", **Shape**: Segmented, Oval, **Color**: Brown or Black, **Legs**: 6, **Wings**: Varies, **Antenna**: Yes.

This ant gets its name from the strong, rotten coconut-like smells it gives off when crushed and the fact that they commonly nest in or around houses. Native to the United States, these ants are very social, living in colonies of up to 100,000 members.

Diet

Odorous house ants like to eat dead insects and sugary sweets, especially melon.

Habitat

Typically living for several years, these ants commonly make their homes in exposed soil, under stones, logs, mulch, debris and other items. They will also nest in wall and floor cracks.

Impact

Odorous House ants do not pose a health threat, but they can contaminate food by leaving waste behind.

Prevention

- Avoid using other individuals' combs, hats, towels, etc.
- Eliminate standing water. Pests, such as ants, mosquitoes and termites, are attracted to moisture.
- Keep tree branches and other plants cut back from the house. Sometimes pests use these branches to get into your home.
- Make sure that there are no cracks or little openings around the bottom of your house. Sometimes pests use these to get into your home.
- Make sure that firewood and building materials are not stored next to your home. Pests like to build nests in stacks of wood.

PAVEMENT ANTS

Common Name: Pavement Ant, **Kingdom**: Animalia, **Phylum**: Arthropoda, **Class**: Insecta, **Order**: Hymenoptera, **Family**: Formicidae, **Species**: *Tetramorium caespitum.*

Size: 1/8", **Shape**: Segmented, Oval, **Color**: Dark Brown to Black, **Legs**: 6, **Wings**: Varies, **Antenna**: Yes.

Although these ants can live inside, they get their name because they make their nests in or under cracks in pavement. They are typically found in the eastern half of the United States, California and Washington. Pavement ant colonies average 3,000 to 4,000 members and have several queens.

Diet

These ants will eat almost anything, including insects, grease, seeds, honeydew, honey, bread, meats, nuts and cheese.

Habitat

This ant gets its name because it most commonly nests in soil next to and beneath slabs, sidewalks, patios, and driveways. Indoors, pavement ants nest under a building's foundation and within hollow foundation walls.

Impact

Pavement ants do not pose a health threat, but they can contaminate food by leaving waste behind.

Prevention

- Eliminate standing water. Pests, such as ants, mosquitoes and termites, are attracted to moisture.
- Keep tree branches and other plants cut back from the house. Sometimes pests use these branches to get into your home.
- Make sure that there are no cracks or little openings around the bottom of your house. Sometimes pests use these to get into your home.
- Make sure that firewood and building materials are not stored next to your home. Pests like to build nests in stacks of wood.

RED IMPORTED FIRE ANTS

Common Name: Red Imported Fire, **Kingdom**: Animalia, **Phylum**: Arthropoda, **Class**: Insecta, **Order**: Hymenoptera, **Family**: Formicidae, **Species**: *Solenopsis invicta*

Size: 1/8" to 3/8", **Shape**: Segmented, Oval, **Color**: Dark Reddish Brown, **Legs**: 6, **Wings**: Varies, **Antenna**: Yes.

Red Imported Fire Ants are more aggressive than other ant species and have a painful sting. These ants and their telltale mound nests should be actively avoided. Red Imported Fire Ants can adapt to many climates and conditions in and around their environment. For example, if the colony senses increased water levels in their nests, they will come together and form a huge ball or raft that is able to float on the water!

Diet

They primarily feed on vegetation.

Habitat

Red imported fire ants will build their nests in mounds of soil outdoors, in landscape areas or near a building's foundation. They occasionally enter buildings through holes or cracks in walls and foundations.

Impact

The sting of a Red Imported Fire Ant is painful and often results in a raised welt that becomes a white blister. Persons allergic to insect stings will react more severely. They are frustrating, not only because of the physical pain they can inflict, but because their mound-building activity can damage plant roots and lead to loss of crops.

Prevention

- Eliminate standing water. Pests, such as ants, mosquitoes and termites, are attracted to moisture.
- Keep tree branches and other plants cut back from the house. Sometimes pests use these branches to get into your home.

- Make sure that there are no cracks or little openings around the bottom of your house. Sometimes pests use these to get into your home.
- Make sure that firewood and building materials are not stored next to your home. Pests like to build nests in stacks of wood.

Antlion

Kingdom: Animalia, **Phylum**: Arthropoda, **Class**: Insecta, **Order**: Neuroptera, **Suborder**: Myrmeleontiformia, **Superfamily**: Myrmeleontoidea, **Family**: *Myrmeleontidae*

Antlions are a group of insects in the family Myrmeleontidae. The most well-known genus is *Myrmeleo*. There are about 2,000 species. Strictly speaking, the term "antlion" applies to the larval form of the members of this family, but while several languages have their own terms for the adult, there is no widely used word for them in English. Very rarely, the adults are called "antlion lacewings". The antlion larva is often called "doodlebug" in North America because of the odd winding, spiralling trails it leaves in the sand while looking for a good location to build its trap, as these trails look like someone has doodled in the sand.

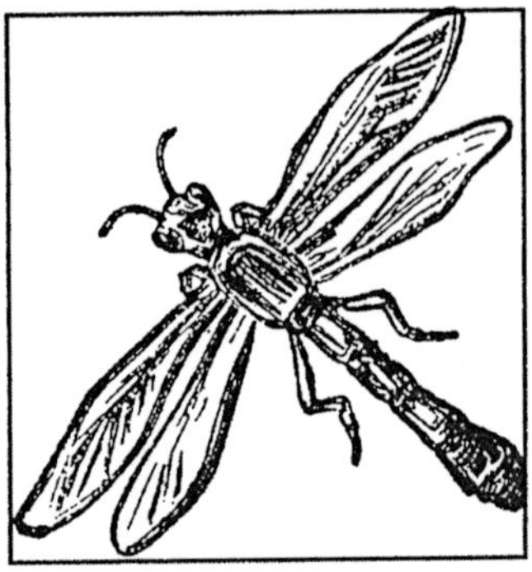

Antlion, common name for an insect larva that digs a funnel-shaped pitfall in fine, loose sand and lies in wait at the bottom for its prey, which consists of ants and other insects that tumble into the pitfall. Hundreds of antlion species are distributed throughout the tropical regions of the world. More than 50 North American species are known, some of which are common in the southern United States. A full-grown antlion is usually about 1.0 cm (about 0.4 in) long and has large, powerful jaws. It pupates in a sand-covered cocoon during the spring; the adult emerges from the cocoon after metamorphosis. It is a delicate-winged, usually gray or brown insect that resembles a damselfly. Antlions are also known as doodlebugs. Antlions make up the family Myrmeleontidae of the order Neuroptera.

Description and Ecology

Antlions are worldwide in distribution, most common in arid and sandy habitats. A few species occur in cold-temperate places; a famous example is the European *Euroleon nostras*, whose scientific name means "our European [ant]lion". They can be fairly small to very large Neuroptera (wingspan range of 2–15 cm). The antlion larvae eat small arthropods - mainly ants, while the adults of some species eat small pollen and nectar, while others are predators of small arthropods in the adult stage too. In certain species of Myrmeleontidae, such as *Dendroleon pantheormis*, the larva, although resembling that of *Myrmeleon* structurally, makes no pitfall, but seizes passing prey from any nook or crevice in which it shelters.

The adult has two pairs of long, narrow, multi-veined wings in which the apical veins enclose regular oblong spaces, and a long, slender abdomen. Although they greatly resemble dragonflies or damselflies, they belong to an entirely different infraclass among the winged insects.

Antlions are easily distinguished from damselflies by their prominent, apically clubbed antennae which are about as long as head and thorax combined. Also, the pattern of wing venation differs, with the very long hypostigmatic cell (behind the fusion point of Sc and R1) being several times as long as wide.

They also are very feeble fliers and are normally found fluttering about in the night, in search of a mate. The adult is thus rarely seen in the wild because it is typically active only in the evening. They are highly active in desert regions and are a nuisance. They will deliver a small, mildly painful bite if given the chance to land on someone.

Life Cycle

The life cycle of the antlion begins with oviposition (egg-laying). The female antlion repeatedly taps the sand surface with the tip of her abdomen. She then inserts her abdomen into the sand and lays an egg. The antlion larva is a ferocious-appearing creature with a robust, fusiform body, a very plump abdomen, the thorax bearing three pairs of walking legs. The prothorax forms a slender mobile "neck" for the large, square, flattened head, which bears an enormous pair of sicklelike jaws with several sharp, hollow projections. The jaws are formed by the maxillae and mandibles, which in each pincer enclose a canal for injecting venom between them. Depending on species and where it lives, the larvae will either hide under leaves or pieces of wood, in cracks of rocks, or dig pits in sandy areas. Antlion larvae are unusual among the insects as they lack an anus. All the metabolic waste that is generated during the larval stage is stored and is eventually emitted as meconium near the end of its pupal stage.[3]

The pupal stage of the antlion is quiescent. The larva makes a globular cocoon of sand stuck together with fine silk spun from a slender spinneret at the posterior end of the body. These cocoons may be buried several centimeters deep in the sand. It remains there for one month, until the completion of the transformation into the sexually mature insect, which then emerges from the case, leaving the pupal integument behind, and climbs to the surface. After about 20 minutes the adult's wings are fully opened and it will fly off in search of a mate. The adult is considerably larger than the larva; they exhibit the greatest disparity in size between larva and adult of any type of holometabolous insects, by virtue of the adults having an extremely thin, flimsy exoskeleton - in other words, they have extremely low mass per unit of volume.

Aphid

Kingdom: Animalia, **Phylum**: Arthropoda **Class**: Insecta, **Order**: Hemiptera, **Suborder**: Sternorrhyncha, **Superfamily**: *Aphidoidea* Latreille, 1802

Aphids constitute the family Aphididae in the order Homoptera.

Aphids, also known as plant lice and in Britain and the Commonwealth as greenflies, blackflies or whiteflies, are small sap sucking insects, and members of the superfamily Aphidoidea. Aphids are among the most destructive insect pests on cultivated plants in temperate regions.

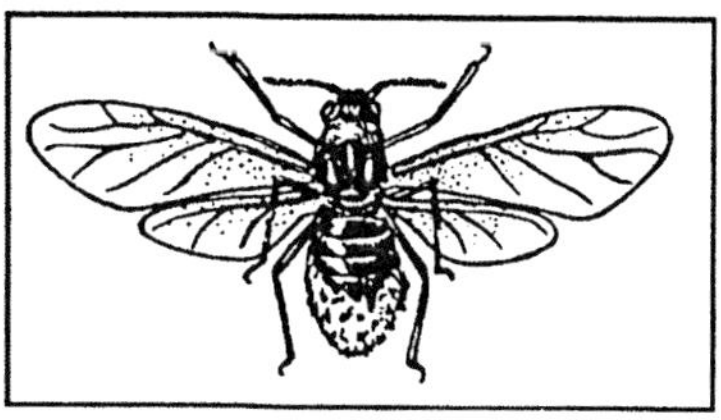

The damage they do to plants has made them enemies of farmers and gardeners the world over, but from a zoological standpoint they are a very successful group of organisms. Their success is in part due to the asexual reproduction capability of some species. About 4,400 species of 10 families are known. Historically, many fewer families were recognised, as most species were included in the family Aphididae. Around 250 species are serious pests for agriculture and forestry as well as an annoyance for gardeners. They vary in length from 1 to 10 millimetres (0.04 to 0.39 in). Natural enemies include predatory ladybirds, hoverfly larvae, parasitic wasps, aphid midge larvae, crab spiders, lacewings and entomopathogenic fungi like *Lecanicillium lecanii* and the Entomophthorales.

The mouthparts of aphids are adapted for piercing and sucking plants and consist of four long, sharp stylets within a proboscis (sheath). Two projecting cornicles, or horn-shaped tubes, at the posterior end of the body emit a waxy secretion. The legs are long and slender but are not adapted for rapid movement; aphids rarely wander far from their birthplace, except during the spring and autumn migratory flights of the females. In autumn the females lay fertilized eggs that survive the winter in crevices and hatch in the spring, producing wingless females that reproduce parthenogenetically (without fertilization from males). The time of development is so short that the eggs sometimes hatch before they are laid. After several generations, winged females are produced; they then migrate to other plants and continue asexual reproduction of wingless females. Toward the end of summer, winged males are produced, which fertilize the winter eggs.

The presence of aphids is readily detected by marks, wrinkles, or abnormalities on the infested plant. Aphids emit from their anus a sweet glutinous substance called honeydew, which is eagerly sought by ants and other insects; ants may hide aphids and protect them from predators. Aphids that feed on such plants as turnip, cabbage, potato, bean, apple, pear, and larch have been the cause of widespread damage, often as a result of transmitting plant viruses. Birds, spiders, parasitic wasps, hover-fly larvae, and lady beetles destroy great numbers of aphids.

Reproduction

Some aphid species have unusual and complex reproductive adaptations, while others have fairly simple reproduction. Adaptations include having both sexual and asexual reproduction, creation of eggs or live nymphs and switches between woody and herbaceous types of host plant at different times of the year. Many aphids undergo cyclical parthenogenesis. In the spring and summer, mostly or only females are present in the population. The overwintering eggs that hatch in the spring result in females, called *fundatrices*.

Reproduction is typically parthenogenetic and viviparous. Females undergo a modified meiosis that results in eggs that are genetically identical to their mother (parthenogenetic). The embryos develop within the mothers' ovarioles, which then give live birth to first instar female nymphs (viviparous).

The offspring resemble their parent in every way except size, and are called *virginoparae*. This process iterates throughout the summer, producing multiple generations that typically live 20 to 40 days. Thus one female hatched in spring may produce many billions of descendants. For example, some species of cabbage aphids (like *Brevicoryne brassicae*) can produce up to 41 generations of females. In autumn, aphids undergo sexual, oviparous reproduction.

A change in photoperiod and temperature, or perhaps a lower food quantity or quality, causes females to parthenogenetically produce sexual females and males. The males are genetically identical to their mothers except that they have one less sex chromosome. These sexual aphids may lack wings or even mouthparts. Sexual females and males mate, and females lay eggs that develop outside the mother. The eggs endure the winter and emerge as winged or wingless females the following spring. This is, for example, the life cycle of the rose aphid (*Macrosiphum rosae*, or less commonly *Aphis rosae*), which may be considered typical of the family. However in warm environments, such as in the tropics or in a greenhouse, aphids may go on reproducing asexually for many years. Some species produce winged females in the summer, sometimes in response to low food quality or quantity.

The winged females migrate to start new colonies on a new plant, often of quite a different kind. For example, the apple aphid (*Aphis pomi*), after producing many generations of wingless females on its typical food-plant, gives rise to winged forms which fly away and settle on grass or corn-stalks. Some aphids have telescoping generations. That is, the parthenogenetic, viviparous female has a daughter within her, who is already parthenogenetically producing her own daughter. Thus a female's diet can affect the body size and birth rate of more than two generations (daughters and granddaughters).

Distribution

Aphids are distributed worldwide, but are most common in temperate zones. Also, in contrast to many taxa, species diversity is much lower in the tropics than in the temperate zones. They can migrate great distances, mainly through passive dispersal by riding on winds. For example, the currant lettuce aphid, *Nasonovia ribisnigri,* is believed to have spread from New Zealand to Tasmania in this way. Aphids have also been spread by human transportation of infested plant materials.

Bee

Kingdom: Animalia, **Phylum**: Arthropoda, **Class**: Insecta, **Order**: Hymenoptera, **Suborder**: Apocrita, **Superfamily**: Apoidea, (**unranked**): *Anthophila.*

Bees are flying insects closely related to wasps and ants, and are known for their role in pollination and for producing honey and beeswax. Bees are a monophyletic lineage within the superfamily Apoidea, presently classified by the unranked taxon name Anthophila. There are nearly 20,000 known species of bees in seven to nine recognized families, though many are undescribed and the actual number is probably higher. They are found on every continent except Antarctica, in every habitat on the planet that contains insect-pollinated flowering plants. Bees are adapted for feeding on nectar and pollen, the former primarily as an energy source and the latter primarily for protein and other nutrients. Most pollen is used as food for larvae. Bees have a long proboscis (a complex "tongue") that enables them to obtain the nectar from flowers. They have antennae almost universally made up of 13 segments in males and 12 in females, as is typical for the superfamily.

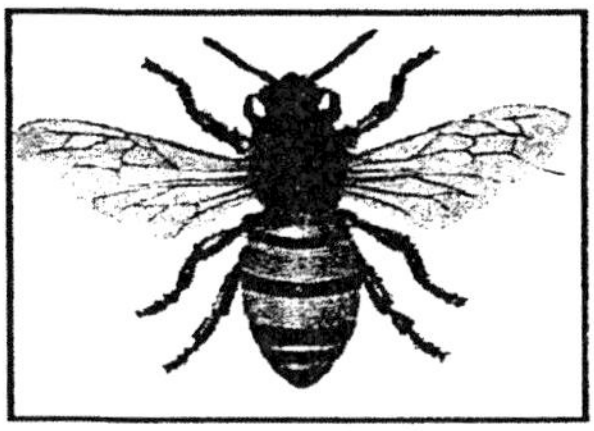

Bees all have two pairs of wings, the hind pair being the smaller of the two; in a very few species, one sex or caste has relatively short wings that make flight difficult or impossible, but none are wingless. The smallest bee is *Trigona minima*, a stingless bee whose workers are about 2.1 mm (5/64") long. The largest bee in the world is *Megachile pluto*, a leafcutter bee whose females can attain a length of 39 mm (1.5"). Members of the family Halictidae, or sweat bees, are the most common type of bee in the Northern Hemisphere, though they are small and often mistaken for wasps or flies. The best-known bee species is the European honey bee, which, as its name suggests, produces honey, as do a few other types of bee. Human management of this species is known as beekeeping or apiculture. Bees are the favourite meal of *Merops apiaster*, the bee-eater bird.

Other common predators are kingbirds, mockingbirds, beewolves, and dragonflies. Bees are dependent on pollen as a protein source and on flower nectar or oils as an energy source. Adult females collect pollen primarily to feed their larvae. Bees are, in fact, the most important pollinating insects, and their interdependence with plants makes them an excellent example of the type of symbiosis known as mutualism, an association between unlike organisms that is beneficial to both parties. The pollen they inevitably lose in going from flower to flower is important to plants because some pollen lands on the pistils (reproductive structures) of other flowers of the same species, resulting in cross-pollination. Most bees have specialized branched or feathery body hairs that help in the collection of pollen. Female bees, like many other hymenopterans, have a defensive sting. Some bees produce honey from flower nectar.

Honey bees and stingless bees commonly hoard large quantities of honey-a characteristic that is exploited by beekeepers, who harvest the honey for human consumption. There are about 20,000 species of bees worldwide. Some species may not yet have been discovered, and many are either not named or have not been well studied. Bees are found throughout the world except at the highest altitudes, in polar regions, and on some small oceanic islands. The greatest diversity of bee species is found in warm, arid or semiarid areas, especially in the American Southwest and Mexico. Bees range in size from tiny species only 2 mm (0.08 in) in length to rather large insects up to 4 cm (1.6 in) long. Many bees are black or gray, but others are bright yellow, red, or metallic green or blue.

Habits and Social Structure

Bees have diverse nesting and social habits. This diversity has provided scientists with a natural laboratory for the study of evolution and social behavior in insects.

SOLITARY AND COMMUNAL BEES

Most other bees, including familiar species of bee such as the Eastern carpenter bee (*Xylocopa virginica*), alfalfa leafcutter bee (*Megachile rotundata*), orchard mason bee (*Osmia lignaria*) and the hornfaced bee (*Osmia cornifrons*) are solitary in the sense that every female is fertile, and typically inhabits a nest she constructs herself. There are no *worker* bees for these species. Solitary bees typically produce neither honey nor beeswax. They are immune from acarine and *Varroa* mites, but have their own unique parasites, pests and diseases. Solitary bees are important pollinators, and pollen is gathered for provisioning the nest with food for their brood. Often it is mixed with nectar to form a paste-like consistency. Some solitary bees have very advanced types of pollen-carrying structures on their bodies. A very few species of solitary bees are being increasingly cultured for commercial pollination. Solitary bees are often oligoleges, in that they only gather pollen from one or a few species/genera of plants (unlike honey bees and bumblebees which are generalists).

No known bees are nectar specialists; many oligolectic bees will visit multiple plants for nectar, but there are no bees which visit only one plant for nectar while also gathering pollen from many different sources. Specialist pollinators also include bee species which gather floral oils instead of pollen, and male orchid bees, which gather aromatic compounds from orchids (one of the only cases where male bees are effective pollinators). In a very few cases only one species of bee can effectively pollinate a plant species, and some plants are endangered at least in part because their pollinator is dying off. There is, however, a pronounced tendency for oligolectic bees to be associated with common, widespread plants which are visited by multiple pollinators (e.g., there are some 40 oligoleges associated with creosote bush in the US desert southwest, and a similar pattern is seen in sunflowers, asters, mesquite, etc.)

Solitary bees create nests in hollow reeds or twigs, holes in wood, or, most commonly, in tunnels in the ground. The female typically creates a compartment (a "cell") with an egg and some provisions for the resulting larva, then seals it off. A nest may consist of numerous cells. When the nest is in wood, usually the last (those closer to the entrance) contain eggs that will become males. The adult does not provide care for the brood once the egg is laid, and usually dies after making one or more nests. The males typically emerge first and are ready for mating when the females emerge.

Providing nest boxes for solitary bees is increasingly popular for gardeners. Solitary bees are either stingless or very unlikely to sting (only in self defense, if ever). While solitary females each make individual nests, some species are gregarious, preferring to make nests near others of the same species, giving the appearance to the casual observer that they are social. Large groups of solitary bee nests are called *aggregations*, to distinguish them from colonies. In some species, multiple females share a common nest, but each makes and provisions her own cells independently. This type of group is called "communal" and is not uncommon.

The primary advantage appears to be that a nest entrance is easier to defend from predators and parasites when there are multiple females using that same entrance on a regular basis The primitive bees, like their relatives the wasps, are solitary. Each female makes her own burrow, in which she constructs earthen chambers to contain her young. She deposits pollen moistened with nectar or oil into individual cells until enough food has accumulated to provide for the young bee from egg hatching until the larva reaches full size. She then lays an egg on the pollen mass and seals the cell before going on to construct another cell.

FAMILIES

There are 11 families of bees. Scientists distinguish between them by subtle differences in wing veins and by the fine structure of the mouthparts and other microscopic characteristics. However, the bees in each family have other interesting descriptive features, including their size, nesting and foraging behaviors, and easy-to-see body features such as body hair, the length of the tongue, and the form of the pollen-carrying structure. Cellophane bees are relatively hairless bees with short, forked tongues. They resemble wasps more so than other bees; hence they are considered the most primitive bees. They line their nest tunnels and larval cells (chambers for the young) with a secretion that hardens into a cellophane-like membrane. They carry pollen on leg hairs or internally in a stomachlike crop. Mining bees are a large group of bees that make soil nests of many branching chambers, each ending in one or more cells. They are either solitary or communal, living in separate but nearby nests. They carry pollen on body and leg hairs.

Sweat bees are generally small, dark-coloured bees with little hair. They, too, usually nest in the ground but may live in societies in which related individuals help each other. Pollen is carried on brushy areas near the base of the legs and on body hairs. Leafcutter bees and mason bees belong to a family of long-tongued bees that have specialized pollen-carrying hairs on the underside of the abdomen. They often make their nests in preexisting cavities and may live in groups of individual nests. Some are used in agriculture to pollinate crops. Digger bees are fast-flying, fuzzy bees that may nest in the ground solitarily or in dense clusters or may excavate nests in wood. They have long tongues and are excellent pollinators of many plants. They carry their pollen on brushy areas near the middle of the hind leg. Carpenter bees are also in the digger bee family.

The most familiar bees are the honey bees and their close relatives. In this family are bees that make intricate nests and live in complex societies. The pollen-carrying structure in these bees is a smooth, bristle-surrounded area on one segment of the hind leg. This structure is known as a pollen basket, or corbicula. This group is divided into four tribes: the orchid bees, the bumble bees, the stingless bees, and the honey bees.

Bees make up a superfamily known as the Apoidea. Cellophane bees make up the family Colletidae, mining bees make up the family Andrenidae, sweat bees make up the family Halictidae, the leafcutter and mason bees and their relatives make up the family Megachilidae, the digger bees make up the family Anthophoridae, and honey bees and their relatives make up the family Apidae.

AFRICANIZED HONEY BEE

Africanized bees, also called killer bees, are a hybrid strain of *Apis mellifera* derived from experiments by Warwick Estevam Kerr to cross European and African honey bees. Several queen bees escaped from his laboratory in South America and have spread throughout the Americas. Africanized honey bees are more defensive than European honey bees.

BUMBLEBEES

Bumblebees (*Bombus terrestris, B. pratorum*) are eusocial in a manner quite similar to the eusocial Vespidae such as hornets. The queen initiates a nest on her own (unlike queens of honey bees and stingless bees which start nests via swarms in the company of a large worker force). Bumblebee colonies typically have from 50 to 200 bees at peak population, which occurs in mid to late summer. Nest architecture is simple, limited by the size of the nest cavity (pre-existing), and colonies are rarely perennial. Bumblebee queens sometimes seek winter safety in honey bee hives, where they are sometimes found dead in the spring by beekeepers, presumably stung to death by the honey bees. It is unknown whether any survive winter in such an environment. Bumblebees are one of the more important wild pollinators, but have declined significantly in recent decades.

CLEPTOPARASITIC BEES

Cleptoparasitic bees, commonly called "cuckoo bees" because their behavior is similar to cuckoo birds, occur in several bee families, though the name is technically best applied to the apid subfamily Nomadinae. Females of these bees lack pollen collecting structures (the scopa) and do not construct their own nests. They typically enter the nests of pollen collecting species, and lay their eggs in cells provisioned by the host bee. When the cuckoo bee larva hatches it consumes the host larva's pollen ball, and if the female cleptoparasite has not already done so, kills and eats the host larva. In a few cases where the hosts are social species, the cleptoparasite remains in the host nest and lays many eggs, sometimes even killing the host queen and replacing her. Many cleptoparasitic bees are closely related to, and resemble, their hosts in looks and size, (i.e., the *Bombus* subgenus *Psithyrus*, which are parasitic bumblebees that infiltrate nests of species in other subgenera of *Bombus*). This common pattern gave rise to the ecological principle known as "Emery's Rule". Others parasitize bees in different families, like *Townsendiella*, a nomadine apid, one species of which is a cleptoparasite of the dasypodaid genus *Hesperapis*, while the other species in the same genus attack halictid bees.

HONEY BEES

Honey bees (or honeybees) are a subset of bees in the genus *Apis*, primarily distinguished by the production and storage of honey and the construction of perennial, colonial nests out of wax. Honey bees are the only extant members of the tribe Apini, all in the genus *Apis*. Currently, there are only seven recognised species of honey bee with a total of 44 subspecies, though historically, anywhere from six to eleven species have been recognised. Honey bees represent only a small fraction of the approximately 20,000 known species of bees. Some other types of related bees produce and store honey, but only members of the genus *Apis* are true honey bees.

Pollination

Species of *Apis* are generalist floral visitors, and will pollinate a large variety of plants, but by no means all plants. Of all the honey bee species, only Apis mellifera has been used extensively for commercial pollination of crops and other plants. The value of these pollination services is commonly measured in the billions of dollars.

Honey

Honey is the complex substance made when the nectar and sweet deposits from plants and trees are gathered, modified and stored in the honeycomb by honey bees as a food source for the colony. All living species of *Apis* have had their honey gathered by indigenous peoples for consumption, though for commercial purposes only *Apis mellifera* and *Apis cerana* have been exploited to any degree. Honey is sometimes also gathered by humans from the nests of various stingless bees.

Beeswax

Worker bees of a certain age will secrete beeswax from a series of glands on their abdomens. They use the wax to form the walls and caps of the comb. As with honey, beeswax is gathered for various purposes.

Pollen

Bees collect pollen in the pollen basket and carry it back to the hive. In the hive, pollen is used as a protein source necessary during brood-rearing. In certain environments, excess pollen can be collected from the hives of *A. mellifera* and *A. cerana*. It is often eaten as a health supplement.

Propolis

Propolis (or bee glue) is created from resins, balsams and tree saps. Those species of honey bees which nest in tree cavities use propolis to seal cracks in the hive. Dwarf honey bees use propolis to defend against ants by coating the branch from which their nest is suspended to create a sticky moat. Propolis is consumed by humans as a health supplement in various ways and also used in some cosmetics.

NOCTURNAL BEES

Four bee families (Andrenidae, Colletidae, Halictidae, and Apidae) contain some species that are crepuscular (these may be either the vespertine or matinal type). These bees have greatly enlarged ocelli, which are extremely sensitive to light and dark; though incapable of forming images. Many are pollinators of flowers that themselves are crepuscular, such as evening primroses, and some live in desert habitats where daytime temperatures are extremely high.

PARASITIC BEES

Parasitic, or cuckoo, bees are those that do not forage or make nests themselves but use the nests and food of other species of bees to provide for their parasitic young. Parasitic bees are of two types: cleptoparasitic bees and social parasites. Cleptoparasitic bees invade the nests of solitary bees, hide their eggs in the brood chambers before the hosts lay theirs, and close the chambers. The young of the parasitic bees then feed on the food that was stored in the chamber by the host female. The eggs or young larvae of the host bee are killed either by the parasitic female or by her larvae. Social parasites are bees that kill the resident queen, lay their own eggs in the host's cells, and then force the host's workers to raise the young parasitic bees. Females of parasitic bees lack such special features as pollen baskets or pollen brushes since they do not forage for food for their young.

SOCIAL BEES

Some bees are communal. They are like solitary bees except that several females of the same generation use the same nest, each making her own cells for housing her eggs, larvae, and pupae. A few kinds of bees are semisocial-they live in small colonies of two to seven bees of the same generation, one of which is the queen, or principal egg layer; the others are worker bees. About 1000 species of bees live in small colonies consisting of a queen and a few daughter workers. In these colonies, the differences in appearance and behavior between workers and queens are scarcely dis-tinguishable. Such species, called primitively eusocial, form temporary colonies that die out in autumn, and only the fertilized queens survive the winter. Bumble bees are familiar examples. The eusocial, or truly social, bees live in large colonies consisting of females of two overlapping generations: mothers (queens) and daughters (workers). Males play no part in the colony's organization and only mate with the queens. Larvae are fed progressively-that is, cells are opened as necessary or are left open so that workers can tend the larvae. Highly eusocial bees, a few hundred species, form permanent colonies in which the queen and worker castes are markedly different in structure, each specialized for its own activities and unable to survive without the other.

Colonies of eusocial bees are complex, highly coordinated societies. Individual bees may have highly specialized functions within the colony. The tasks of defense, food collection and storage, reproduction, and many other activities are regulated by the colony's response to environmental conditions inside and outside the hive. Individuals communicate by means of chemical messages, touch, sound, and, in the case of honey bees, a symbolic dance language. The nests of many eusocial bees are very elaborate and may be constructed partially of wax secreted by the bees.

STINGLESS BEES

Stingless bees, sometimes called stingless honey bees or simply meliponines, are a large group of bees, comprising the tribe Meliponini (or subtribe Meliponina according other authors). They belong in the family Apidae, and are closely related to common honey bees, carpenter bees, orchid bees and bumblebees.[3] The common name is slightly misleading, as a great many other bee species, especially in the family Andrenidae, are also incapable of stinging, as are all male bees. Meliponines have stingers, but they are highly reduced and cannot be used for defense.

Pollination

Australian farmers rely heavily on the introduced Western honey bee to pollinate their crops. However, for some crops native bees may be better pollinators. Stingless bees have been shown to be valuable pollinators of crops such as macadamias and mangoes. They may also benefit strawberries, watermelons, citrus, avocados, lychees and many others. Research into the use of stingless bees for crop pollination in Australia is still in its very early stages, but these bees show great potential.

Beetle

Kingdom: Animalia, **Phylum**: Arthropoda, **Class**: Insecta **Subclass**: Pterygota, **Infraclass**: Neoptera, **Superorder**: Endopterygota, **Order**: **Coleoptera** Linnaeus, 1758.

Beetles are insects of the order Coleoptera, which contains more species than any other order in the animal kingdom, constituting about 25% of all known life-forms. About 40% of all described insect species are beetles (about 400,000 species), and new species are discovered frequently. Some estimates put the total number of species, described and undescribed, at as high as 100 million, but one million is a more likely figure. The largest taxonomic family, the Curculionidae (the weevils or snout beetles), also belongs to this order. Beetle, common name for any member of an order of insects, the most prominent characteristic of which are the elytra, the hardened, sheathlike front wings, which usually cover the entire abdomen when the insect is not in flight. In some beetles, such as the rove beetles, the elytra are short, covering only a portion of the abdomen. Some beetles are wingless. Beetles vary greatly in size. The largest is the Hercules beetle, which grows to 16 cm (up to 6.3 in) in length. Other species may be less than 0.1 cm (less than 0.04 in) long.

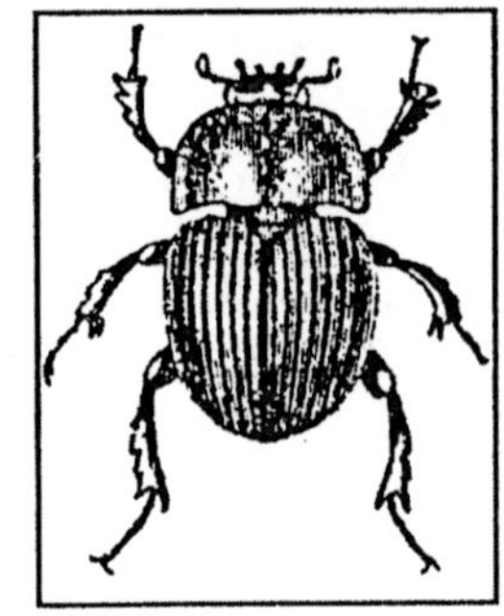

Beetles can be found in almost all habitats, but are not known to occur in the sea or in the polar regions. They interact with their ecosystems in several ways. They often feed on plants and fungi, break down animal and plant debris, and eat other invertebrates. Some species are prey of various animals including birds and mammals. Certain species are agricultural pests, such as the Colorado potato beetle (*Leptinotarsa decemlineata*), the boll weevil (*Anthonomus grandis*), the red flour beetle (*Tribolium castaneum*), and the mungbean or cowpea beetle (*Callosobruchus maculatus*), while other species of beetles are important controls of agricultural pests. For example, beetles in the family Coccinellidae ("ladybirds" or "ladybugs") consume aphids, scale insects, thrips, and other plant-sucking insects that damage crops.

Although they also vary greatly in form, their basic anatomical structures are similar. Beetles undergo complete metamorphosis. The larvae may be sedentary blind grubs or agile predators. Some larvae change their form with each molt. All larvae have three pairs of legs on the thorax. The pupae are usually encased in a thin, light-coloured skin with the legs free. The adults have biting mouth parts, in some cases enormously developed. The beetle order embraces more species than any other group in the animal kingdom. At least 250,000 species are known-more than one-quarter of all animal species. About 160 families exist. Although some families contain only one or two species, others, such as the weevil, contain 30,000 species-far more, for example, than all mammal species. The classification of this enormous number of forms is extremely difficult. The order is generally divided into four suborders, which are, in turn, subdivided into series and superfamilies.

The families are further divided into subfamilies, and these are subdivided into tribes and genera. Some entomologists find still other groupings necessary to indicate the many relationships and differences among beetles. Beetles vary widely in their habits and are found under the most diverse conditions. A few live in salt water, more in fresh water, and a small number breed in hot springs. Some beetles live under the bark of living and dead trees. Numerous beetles feed on the roots, wood, leaves, flowers, and fruit of living plants, causing great economic damage.

Reproduction

Beetles may display extremely intricate behaviour when mating. Pheromone communication is likely to be important in the location of a mate. Conflict can play a part in the mating rituals of species such as burying beetles (genus *Nicrophorus*) where conflicts between males and females rage until only one of each is left, thus ensuring reproduction by the strongest and fittest. Many male beetles are territorial and will fiercely defend their small patch of territory from intruding males. In such species, the males may often have horns on the head and/or thorax, making their overall body lengths greater than those of the females, unlike most insects. Pairing is generally short but in some cases will last for several hours.

During pairing sperm cells are transferred to the female to fertilise the egg. Parental care varies between species, ranging from the simple laying of eggs under a leaf to certain scarab beetles, which construct underground structures complete with a supply of dung to house and feed their young. Other beetles are leaf rollers, biting sections of leaves to cause them to curl inwards, then laying their eggs, thus protected, inside. Beetles make up the order Coleoptera. The Hercules beetle belongs to the family Scarabaeidae. It is classified as Dynastes hercules.

Control

Japanese beetle control is difficult, but altering their habitat can help reduce their population. Growers can avoid irrigation while the eggs and larvae are developing, especially in areas with low natural rainfall. It's also possible to avoid planting preferred species near the kind of soil Japanese beetles prefer for laying. Red and silver maples, dogwood, cedar, lilac and other species are more resistant to these pests.

BARK BEETLE

A bark beetle is one of approximately 220 genera with 6,000 species of beetles in the subfamily Scolytinae. Traditionally this was considered a distinct family Scolytidae, but nowadays it is understood that bark beetles are in fact very specialized members of the "true weevil" family (Curculionidae). Well-known species are members of the type genus *Scolytus* - namely the European elm bark beetle *S. multistriatus* and the large elm bark beetle *S. scolytus*, which like the American elm bark beetle (*Hylurgopinus rufipes*) transmit Dutch elm disease fungi (*Ophiostoma*). The mountain pine beetle (*Dendroctonus ponderosae*), southern pine beetle (*Dendroctonus frontalis*) and their near relatives are major pests of conifer forests in North America. A similarly aggressive species in Europe is the spruce Ips (*Ips typographus*). A tiny bark beetle is the major pest of coffee around the world—the coffee berry borer, (*Hypothenemus hampei*). Most bark beetles are brown and compact in shape. They resemble weevils, their closest relatives, except they lack the prolonged snout.

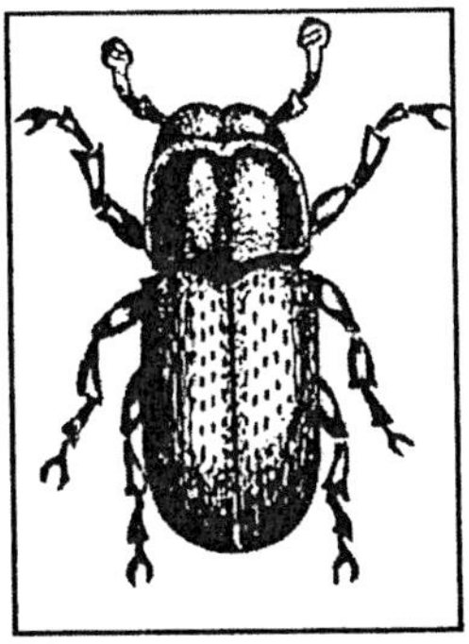

Male bark beetles are attracted to females by a sex pheromone (volatile odour) and mate in or near the female's larval burrow. Females migrate to a new place on the same tree or to a new tree and start another burrow. The female lays her eggs in pockets at intervals along the tunnel. When the eggs hatch as larvae, each individual forms its own tunnel at an angle to the original burrow made by the female. The larval tunnels may radiate outward under the bark to form a gallery, the shape of which is characteristic of certain species. The larvae are white, legless grubs. After their last growth stage, they form a pupal case of the last larval skin and begin to metamorphose into adult beetles. When the beetle emerges from the pupa as an adult, it burrows directly to the surface of the tree. Adults searching for new trees are attracted to the odors given off by damaged or diseased wood, especially those under attack by other bark beetles. Bark beetles kill trees by girdling the sapwood with their burrows, cutting off the flow of nutrients to the tree's roots.

A healthy tree may resist bark beetle attack by flooding the beetle tunnels with its own thick sap, or pitch, under high pressure. Some species of bark beetle also carry plant diseases, especially fungi. The smaller European elm bark beetle, with black thorax and red wing cases, is a principle carrier of Dutch elm disease. The western pine beetle is a major pest of evergreen forests in the western United States. The damage caused by these species and others like them results in the loss of millions of dollars a year in timber resources and is a major concern of agriculture and forestry.

The smaller European elm bark beetle is Scolytus multistriatus and the western pine beetle is Dendroctonus brevicomis Bark beetles make up the family Scolytidae, order Coleoptera.

BLISTER BEETLES

Blister beetles are beetles (Coleoptera) of the family Meloidae, so called for their defensive secretion of a blistering agent, cantharidin. There are approximately 7,500 known species worldwide. Many are conspicuous and some aposemati-cally coloured, announcing their toxicity to would-be predators. The insects of this family pass through a complex metamorphosis, some having as many as nine distinct stages from egg to adult. The early-stage larvae are fast-moving parasites of other insects, while later-stage larvae are parasitic but sedentary.

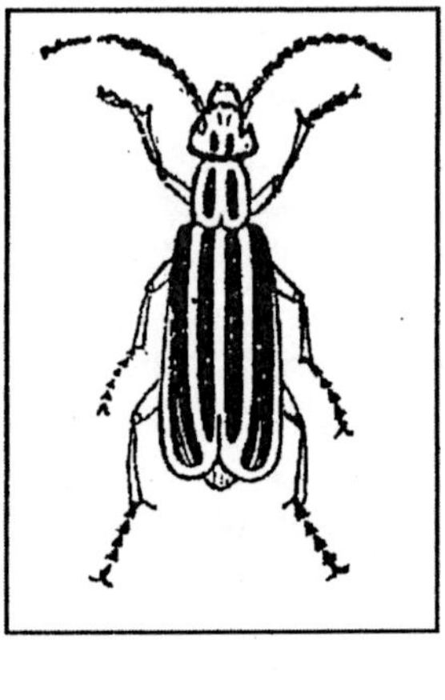

Hosts include wild bee larvae and the buried egg masses of grasshoppers. The adult beetles are vegetarian and some are minor pests of crops. For example, the striped blister beetle, with black and orange stripes and the margined blister beetle, black with gray margins, are what many people think of as old-fashioned potato beetles. The

Spanishfly is another species of blister beetle, famous for the aphrodisiac qualities of its extract. Another group of species are known as oil beetles because they secrete an oily substance from their joints when disturbed.

Blister beetles make up the family Meloidae, order Coleoptera. The striped blister beetle is classified as Epicauta vittata, the margined blister beetle is E. pestifera, and the Spanishfly is Lytta vesicatoria. Oil beetles make up the genus Meloe.

SPANISH FLY

The Spanish fly is an emerald-green beetle in the family Meloidae, *Lytta vesicatoria*. Other species of blister beetle used by apothecaries are often called by the same name. *Lytta vesicatoria* is sometimes incorrectly called *Cantharis vesicatoria*, but the genus *Cantharis* is in an unrelated family, Cantharidae. The beetle contains up to five per cent cantharidin which irritates animal tissues. The crushed powder of Spanish fly is of yellowish brown to brown-olive color with iridescent reflections, of disagreeable scent and bitter flavor. Spanish fly, or cantharides as it is sometimes called, is often given to farm animals to incite them to mate. When ingested, preparations containing cantharides once excreted in the urine irritate the urethral passages, causing inflammation in the genitals and subsequent priapism. For this reason, Spanish fly has been given to humans for purposes of seduction; they help men keep a longer erection. It is dangerous since the amount required is minuscule and the difference between the effective dose and the toxic dose is quite narrow (narrow therapeutic index).

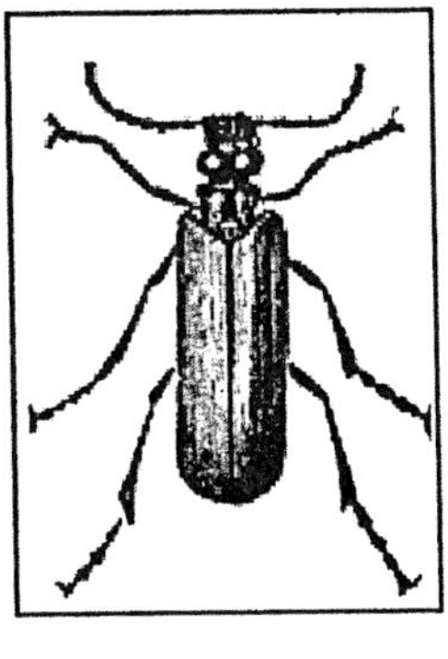

Cantharides cause painful urination, fever, and sometimes bloody discharge. They can cause permanent damage to the kidneys and genitals. The adult European Spanish fly is shiny green, sometimes tinged with reddish yellow. It has a soft, elongated body, 12 to 21 mm (0.47 to 0.88 in) long. Like other blister beetles, the thorax, or neck, is narrow, and the head is commonly wider than the body. They are herbivores, feeding chiefly on pollen and petals of ash, lilac, and privet. Eggs are laid in batches in shallow burrows in soil. The emergent larvae, known as triungulins, are active predators. They seek nests of ground-dwelling bees and kill and consume the bee larvae.

Triungulins have a broad head and thorax, well-developed legs and antennae, and a pair of tail filaments called cerci. Blister-beetle larvae develop by hypermetamor-phosis: Each growth stage, or instar, is different in appearance and habits. The head and appendages of the second through fifth instars become reduced in size with each molt. The abdomen becomes more robust, with indistinct segments, and later stages do not feed. Sixth instars are greatly reduced in size and become inactive pseudopupae, from which adults emerge in spring.

Spanishfly has one generation each year. Spanishfly has been harvested for its extract for over 2000 years. The beetles are shaken onto cloths from the bushes and trees where they feed. Then they are killed in boiling water or vinegar, dried in the sun or in ovens. Cantharidin obtained from the dried, crushed bodies of Spanishfly and mixed into various medicines has been used as a salve, diuretic, and aphrodisiac. The ancient Greeks and Romans also used it as a poison. Cantharidin is still used today as an ingredient in remedies for urogenital disorders. Its benefits as a sex stimulant have been largely discredited. Some other species of blister beetles in the Spanishfly genus are pests because they eat crops or their bodies contaminate harvested fodder. For example, adults of the Nuttall blister beetle feed on alfalfa and other legumes; their bodies commonly poison horses and other livestock that eat the beetle-infested crop.

The European Spanishfly is classified as *Lytta vesicatoria*; the Nuttall blister beetle is classified as *Lytta nuttalli. Lytta vesicatoria* is 15 mm to 22 mm long and 5 mm to 8 mm wide.

Spanishfly beetles belong to the genus *Lytta* of the family Meloidae.

BOLL WEEVIL

The boll weevil (*Anthonomus grandis*) is a beetle measuring an average length of six millimetres, which feeds on cotton buds and flowers. Thought to be native to Central America, it migrated into the United States from Mexico in the late 19th century and had infested all U.S. cotton-growing areas by the 1920s, devastating the industry and the people working in the American south. During the late 20th century it became a serious pest in South America as well. Since 1978, the Boll Weevil Eradication Program in the U.S. has allowed full-scale cultivation to resume in many regions. The adult insect has a long snout, is grayish in color, and is usually less than 6 mm (less than 0.24 in) long. Feeding only on the cotton plant, it begins in early spring to puncture the buds and bolls and lay its eggs in them.

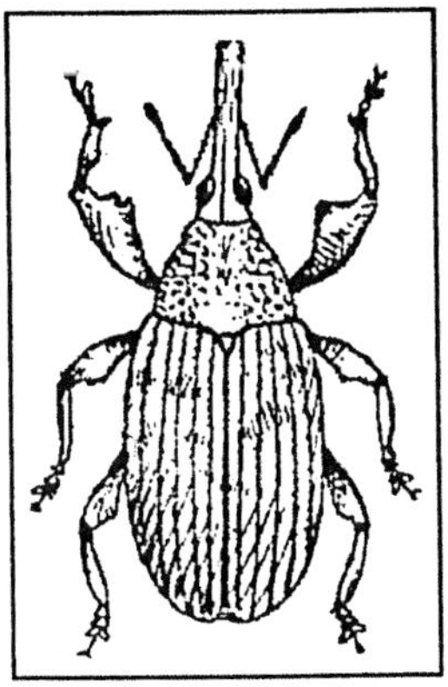

The eggs hatch into larvae in three to five days. The life cycle of the boll weevil from egg to egg-laying adult is about three weeks. Four or five generations may breed in one season. The larva, a fat, white maggot, does the most damage. It lives on the internal tissues of the buds and bolls. An infested bud usually drops, but most of the damaged bolls remain on the plant and become stunted or dwarfed. Adult weevils that emerge in the autumn hibernate in grass, old bolls, or other vegetation or in the seeds around the cotton gins. They reappear in spring.

Life Cycle

Adult weevils overwinter in well-drained areas in or near cotton fields after diapause. They emerge and enter cotton fields from early spring through midsummer, with peak emergence in late spring, and feed on immature cotton bolls. The female lays about 200 eggs over a 10-12 day period. The oviposition leaves wounds on the exterior of the flower bud. The eggs hatch in three to five days. The larvae feed within the cotton squares for eight to ten days, then pupate. The pupal stage lasts five to seven days. The life cycle from egg to adult spans about three weeks during the summer. Under optimal conditions there may be eight to 10 generations per season. According to the book "From Can See to Can't" by Thad Sitton and Dan Utley, "Under ideal conditions for reproduction—which fortunately seldom existed—the progeny of a single pair of weevils emerging in the spring could reach something like 134 million before the coming of frost." Boll weevils will begin to die at temperatures at or below 23 degrees Fahrenheit. Research at the University of Missouri indicates they cannot survive more than an hour at five degrees Fahrenheit.

The insulation offered by leaf litter, crop residues, and snow may enable the beetle to survive when air temperatures drop to these levels. Other limitations on boll weevil populations include extreme heat and drought. Its natural predators include fire ants, insects, spiders, birds, and a parasitic wasp, *Catolaccus grandis*. The insects at times engage in what seems to be almost suicidal behavior by emerging from diapause before cotton buds are available.

The boll weevil belongs to the family Curculiondae, order Coleoptera. It is classified as Anthonomus grandis.

BOMBARDIER BEETLE

Bombardier beetles are ground beetles (Carabidae) in the tribes Brachinini, Paussini, Ozaenini, or Metriini—more than 500 species altogether—which are most notable for the defense mechanism that gives them their name: When disturbed, the beetle ejects a noxious chemical spray in a rapid burst of pulses from special glands in its abdomen. The ejection is accompanied with a popping sound. A bombardier beetle produces and

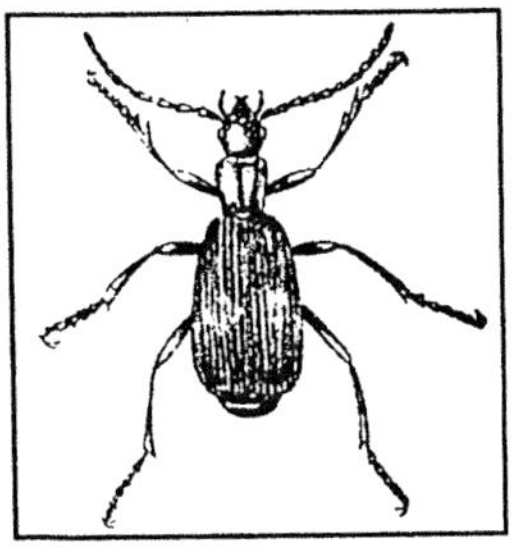

stores two reactant chemical compounds, hydroquinone and hydrogen peroxide, in separate reservoirs in the rear tip of its abdomen. When threatened, the beetle contracts muscles that force the two reactants through valved tubes into a mixing chamber containing water and a mixture of catalytic enzymes. When combined, the reactants undergo a violent exothermic chemical reaction, raising the temperature to near the boiling point of water. The corresponding pressure buildup forces the entrance valves from the reactant storage chambers to close, thus protecting the beetle's internal organs.

The boiling, foul-smelling liquid partially becomes a gas (flash evaporation) and is expelled through an outlet valve into the atmosphere with a loud popping sound. The flow of reactants into the reaction chamber and subsequent ejection to the atmosphere occurs cyclically at a rate of about 500 times per second and with the total pulsation period lasting for only a fraction of a second. The gland openings of some African bombardier beetles can swivel through 270° and thrust between the insect's legs, so it can be discharged in all sorts of directions with considerable accuracy. Bombardier beetles belong to the family Carabidae in the order Coleoptera, class Insecta. The common American species that is 1.3 cm (0.5 in) long is *Brachinus fumans*.

CARPET BEETLE

Carpet beetle, common name for a member of the hide beetle family, the larvae of which feed on carpets and various other materials of animal origin. The adult beetle is about 4 mm (about 0.16 in) long and oval in form, with black, red, and white markings on the back and white scales on the underside. The spindle-shaped larva has tufts of stiff bristles along the sides and at the ends of the body. The beetle eggs are deposited in cracks in the floor under carpet. The developing larvae then

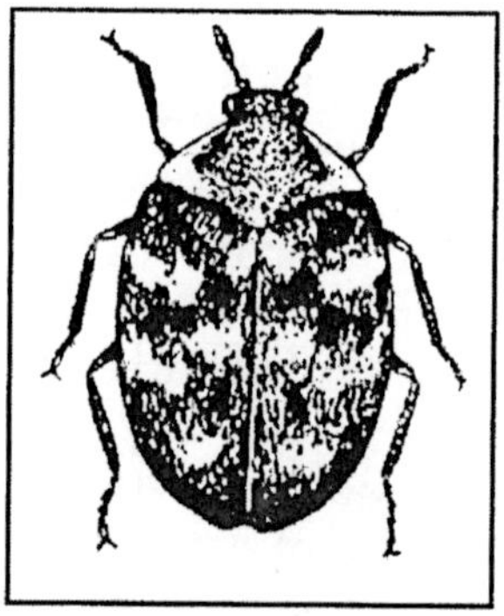

feed on the carpet, often following the cracks and cutting long slits. They are also pests of museum specimens, especially the remains of other insects.

Life Cycle

The eggs hatch between 10 to 35 days depending on the temperature. The larvae can last at least a year depending on the quality of the food supply. The adult beetle lives from 7-41 days. The adults appear in April, May and June and their resulting larvae hibernate during the following winter pupating during the latter part of February and March. On emergence the adult beetles seek the light so they usually fly to the windows. They fly to light coloured flowers where they feed on nectar and pollen. After mating, the females enter houses during late summer and autumn and lay their eggs in birds' nests in roof voids and other suitable places. The larvae feed on feathers and wool soiled with excrement, dead fledglings, etc.

Control

Infestations of these insects mostly originate in birds' nests such as those of house-sparrow, house-martin, swift and starling. The carpet beetle larvae subsist on the soiled wool and feathers in the nest-lining and then manage to get under the eaves in the roof space. From here they crawl into the upper rooms of the house, usually following the hot water pipe system. They appear to be attracted to warmer situations. Airing cupboards often harbour substantial infestations. Carpet beetles do, however, sometimes fly through open windows and presumably initiate infestations directly, but infestations arising from birds' nest cause continual invasions. The first rule of control therefore is to remove birds' nests from eaves and all similar situations. If dead birds or rodents are found in chimneys or under floors they should be removed. Any pieces of old carpet or similar material in lofts or roof voids must be destroyed otherwise they will act as reservoirs of infestation. In trunks or cases, a deterrent application of paradichlorobenzene or napthalene may be effective if the lids are tightly fitting. One pound of crystals of the particular chemical should be used for each ten cubic feet of space. The young larvae and the beetles are the most easily killed by these chemicals but the containers must be kept tightly closed. The carpet beetle belongs to the family Dermestidae, order Coleoptera. It is classified as *Anthrenus scrophulariae.*

CATERPILLAR HUNTER

The forest caterpillar hunter *(Calosoma sycophanta)* is a large, bright green, metallic ground beetle that is native to Europe. In 1905 it was imported to New England for control of the gypsy moth. The species is a voracious consumer of caterpillars during both its larval stage and as an adult. Caterpillar hunters belong to the family Carabidae. The fiery searcher is classified as Calosoma scrutator, and the European caterpillar hunter as *Calosoma sycophanta.*

CLICK BEETLE

The family Elateridae is commonly called click beetles (or "typical click beetles" to distinguish them from the related Cerophytidae and Eucnemidae), elaters, snapping beetles, spring beetles or "skipjacks". They are a cosmopolitan beetle family characterized by the unusual click mechanism they possess. There are a few closely-related families in which a few members have the same mechanism, but all elaterids can click. A spine on the prosternum can be snapped into a corresponding notch on the mesosternum, producing a violent "click" which can bounce the beetle into the air. Clicking is mainly used to avoid predation, although it is also useful when the beetle is on its back and needs to right itself. There are about 9300 known species worldwide, and 965 valid species in North

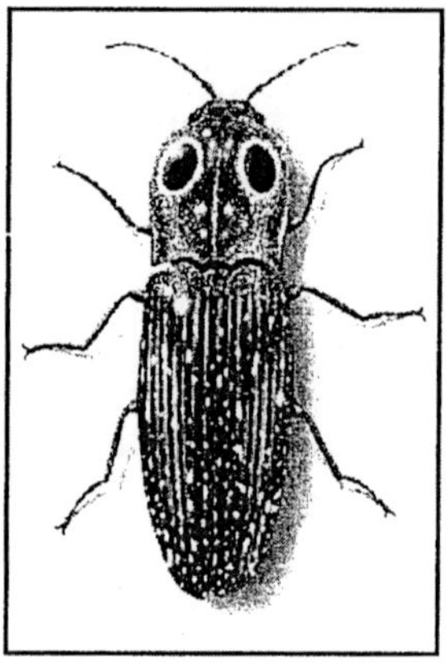

America It bends its head and prothorax backward and then straightens suddenly with a snapping motion, sometimes propelling itself as much as 15 cm (6 in) into the air. The slender, hard-coated larvae of the click beetle, commonly called wireworms, live in the soil or in dead wood. Some species are very destructive to the roots of plants, and fall plowing is considered the most effective way of controlling them. More than 800 species of click beetles are found in North America. The eyed click beetle, also known as the eyed elater, is one of the largest species in the United States. It is dark grayish with two large eyelike spots on the thorax. Some click beetles are bioluminescent and may be mistaken for fireflies. Click beetles make up the family Elateridae, order Coleoptera. The eyed click beetle is classified as *Alaus oculatus*.

Click beetles can be large and colorful (e.g. Selatosomus cruciatus), but most are small to medium-sized (<2 cm) and dull. The adults are typically nocturnal and phytophagous, but rarely of economic importance. In hot weather, they may enter people's houses at night if entries or windows are left opened but are not a pest. Click beetle larvae, called wireworms, are usually saprophagous, but some species are serious agricultural pests, and other species are predators of other insect larvae. A few elaterid species are bioluminescent (in both larvae and adult forms), such as *Pyrophorus*.

Larvae are slender, elongate, cylindrical or somewhat flattened, and relatively hard-shelled for larvae—bearing resemblance to common mealworms (tenebrionid larvae; e.g. *Tenebrio molitor*). The three pairs of legs on the thoracic segments are short and the last abdominal segment is, as is frequently the case in beetle larvae, directed downwards and may serve as a terminal proleg in some species. The posterior end of the body (9th abdominal segment) is pointed in larvae of *Agriotes, Dalopius* and *Melanotus,* but is bifid due to a so-called caudal notch in *Selatosomus* (formerly *Ctenicera*), *Limonius, Hypnoides* and *Athous* species. The dorsum of the 9th abdominal segment may also have sharp processes (e.g. in Oestodini such as *Drapetes* spp. and *Oestodes* spp.).

Although some species complete their development in one year (e.g. *Conoderus*), wireworms usually spend three or four years in the soil, feeding on decaying vegetation and the roots of plants, and often causing damage to agricultural crops such as potato, strawberry, corn, and wheat. The subterranean habits of wireworms, their ability to quickly locate food by following carbon dioxide gradients produced by plant material in the soil, and their remarkable ability to recover from illness induced by insecticide exposure (sometimes after many months), make it hard to exterminate them once they have begun to attack a crop. Wireworms can pass easily through the soil on account of their shape and their propensity for following pre-existing burrows, and can travel from plant to plant, thus injuring the roots of multiple plants within a short time. Other subterranean creatures such as the leatherjacket grub of crane flies which have no legs, and geophilid centipedes, which may have over two hundred, are sometimes confounded with the six-legged wireworms. Larvae of some Brazilian savanna species of click beetles live in burrows scattered over the surfaces of termite mounds using their bioluminescence to attract flying prey.

FIREFLY

Lampyridae is a family of insects in the beetle order Coleoptera. They are winged beetles, and commonly called fireflies or lightning bugs for their conspicuous crepuscular use of bioluminescence to attract mates or prey. Fireflies produce a "cold light", with no infrared or ultraviolet frequencies. This chemically-produced light from the lower abdomen may be yellow, green, or pale-red, with wavelengths from 510 to 670 nanometers.

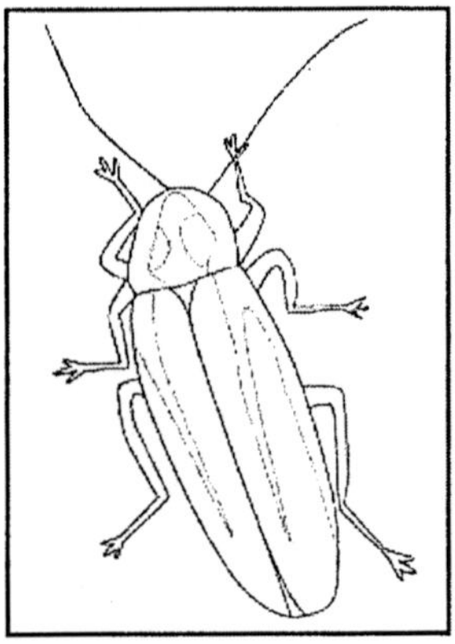

There are 2,000 species of firefly found in temperate and tropical environments. Many are in marshes or in wet, wooded areas where their larvae have abundant sources of food. These larvae emit light and are often called "glowworms", in particular, in Eurasia. In the Americas, "glow worm" also refers to the related Phengodidae. Fireflies, also known as glow worms, tend to be brown and softbodied, often with the elytra more leathery than in other beetles. Though the females of some species are similar in appearance to males, larviform females are found in many other firefly species. These females can often be distinguished from the larvae only because they have compound eyes. The members of one species of firefly in tropical Asia congregate in great numbers in trees and engage in synchronous flashing that begins with a single "pacemaker" insect. Waves of light travel up and down the tree or pass from tree to tree. Usually, only the males take part in this synchronous flashing.

The females of certain unusual insect-eating North American fireflies are known to mimic the flashes of other, nonpredatory females. When an unsuspecting male alights nearby, the predatory female reduces the intensity of her flashes to more closely resemble the weaker signals of the non-predatory female until her prey is within reach. Most adult fireflies, however, eat only pollen and nectar or do not eat at all. The most commonly-known fireflies are nocturnal, though there are numerous species that are diurnal. Most diurnal species are non-luminescent, though some species that remain in shadowy areas can produce light. A few days after mating, a female lays her fertilized eggs on or just below the surface of the ground. The eggs hatch 3–4 weeks later and the larvae feed until the end of the summer.

The larvae are commonly called glowworms, not to be confused with the distinct beetle family Phengodidae or fly genus *Arachnocampa*. Lampyrid larvae have simple eyes. The term glowworm is also used for both adults and larvae of species such as *Lampyris noctiluca*, the common European glowworm, in which only the non-flying adult females glow brightly and the flying males glow only weakly and intermittently. Fireflies hibernate over winter during the larval stage, some species for several years. Some do this by burrowing underground, while others find places on or under the bark of trees. They emerge in the spring. After several weeks of feeding, they pupate for 1 to 2.5 weeks and emerge as adults. The larvae of most species are specialized predators and feed on other larvae, terrestrial snails, and slugs. Some are so specialized that they have grooved mandibles that deliver digestive fluids directly to their prey.

Adult diet varies. Some are predatory, while others feed on plant pollen or nectar. Most fireflies are quite distasteful and sometimes poisonous to vertebrate predators. This is due at least in part to a group of steroid pyrones known as lucibufagins (LBGs), which are similar to cardiotonic bufadienolides found in some poisonous toads. Fireflies make up the family Lampyridae, of the order Coleoptera. The pyralis is classified as Photinus pyralis. The insect-eating North American fireflies are classified in the genus Photuris. Phengodid glow-worms belong to the family Phengodidae.

Life Cycle

The life cycle of most firefly species takes two years. A few days after mating, which occurs in the spring, a female lays her fertilized eggs on or just below the surface of the ground. The eggs hatch 3-4 weeks later and the larva feed until the end of the summer. In the United States, fireflies may persist in the larvae stage for 1-3 years. The larvae are commonly called glowworms, not to be confused with the distinct beetle family Phengodidae. Lampyrid larvae have simple eyes.

Fireflies overwinter (some species for several years) during the larval stage. Some do this by burrowing underground, while others find places on or under the bark of trees. They emerge as adults in the spring. After several weeks of feeding, they pupate for one to 2.5 weeks and emerge as adults. The adult firefly's life span lasts only a few weeks, during which reproduction is their main goal. The larvae of most species are specialized predators and feed on other larvae, small insects, earthworms, terrestrial snails, and slugs. Some are so specialized that they have grooved mandibles, which deliver toxic digestive fluids directly into their prey before sucking out the liquefied body contents. The diet of adults is variable. It has been reported that some are predatory and some feed on plant pollen or nectar.

Habitat

Fireflies live in various habitats. Many species thrive in forests, fields or the margins between them. Some live in more arid areas, but they typically follow the rainy season. Fireflies are found all over the world, from North and South America to Europe and Asia. Most firefly species have one thing in common: standing water. They live near ponds, streams, marshes, rivers and lakes, but they don't need a lot of water to get by. Vernal pools and small depressions that hold water during firefly mating season can all provide the habitat fireflies need. Most firefly species live at the margins where forest or field meet water.

GROUND BEETLE

Ground beetles or carabids are collective terms for the beetle family Carabidae. This is a large family, with more than 40,000 species worldwide, approximate-ly 2,000 of which are found in North America and 2,700 in Europe. Although there is some variation in their body shape and coloring, most are shiny black or metallic and have ridged wing covers (elytra). The elytra are fused in some species, particularly large Carabinae, rendering

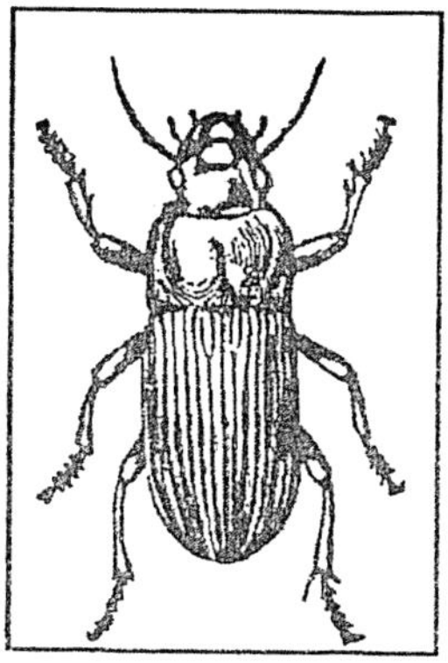

the beetles unable to fly. The genus *Mormolyce* is known as violin beetles due to their peculiarly shaped elytra. All carabids except the quite primitive flanged bombardier beetles (Paussinae) have a groove on their foreleg tibiae bearing a comb of hairs. This is used for cleaning their antennae. Ground beetles are agriculturally important because they destroy such harmful insects as the potato beetle, browntail moth, gypsy moth, cutworm, cankerworm, June beetle, and plum curculio. Few ground beetles are considered harmful; some species feed on seeds, corn, and strawberries. Ground beetles make up the family Carabidae of the order Coleoptera.

Habitat

Common habitats are under the bark of trees, under logs, or among rocks or sand by the edge of ponds and rivers. Most species are carnivorous and actively hunt for any invertebrate prey they can overpower. Some will run swiftly to catch their prey; tiger beetles (Cicindelinae) can sustain speeds of 8 km/h (5 mph) – in relation to their body length they are among the very fastest land animals on Earth. Unlike most Carabidae which are nocturnal, the tiger beetles are active diurnal hunters and often brightly coloured; they have large eyes and hunt by sight. Ground beetles of the species *Promecognathus laevissimus* are specialised predators of the cyanide millipede (*Harpaphe haydeniana*), countering the hydrogen cyanide which makes these millipedes poisonous to most carnivores.

Subfamilies and Selected Genera

The taxonomy used here is based on the *Catalogue of Palaearctic Coleoptera* and the *Carabidae of the World Database*. Other classifications, while generally agreeing with the division into a basal radiation of more primitive lineages and the more advanced group informally called Carabidae Conjunctae, differ in details. For example, the system used by the Tree of Life Web Project makes little use of subfamilies, listing most tribes as *incertae sedis* as to subfamily. *Fauna Europaea* on the other hand splits the Harpalinae instead of lumping them, restricting them to what in the system use here is the tribe Harpalini.

All the approaches mentioned above are legitimate as they agree with the phylogeny as far as it has been resolved. The inclusive Harpalinae presented here are used for two reasons, one scientific and one practical – first, the majority of authors presently uses this system, following the *Catalogue of Palaearctic Coleoptera.*

GRUB

Indeed, it was most likely the larva of a glowworm, properly called a phengodid, family Phengodidae. The California species are in the genus Zarhipis. They feed exclusively on millipedes. The adults are beetles, but only the male resembles a beetle. Adult females retain their larval appearance (so it could have been an adult female you found). Firefly larvae also glow, but only from the tail end. Grub, in entomology, familiar term for the larval stage of certain beetles. Most specifically, the term is used for larvae with soft, thick bodies, well-developed heads, and legs on the thorax but not on the abdomen. In general, grubs tend to be pale in color. They are usually slow-moving and many are soil dwellers. Similar larvae of many other insects are also called grubs.

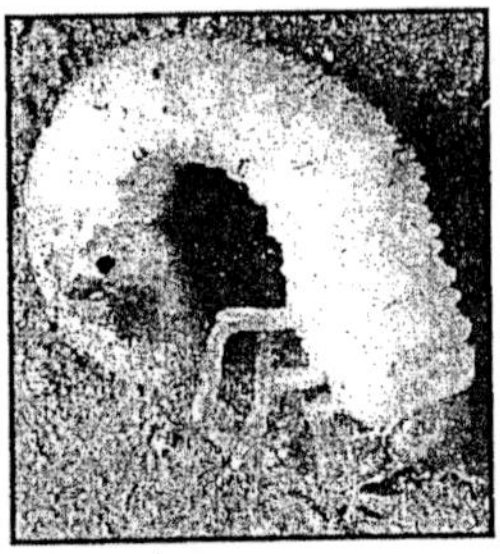

JAPANESE BEETLE

The Japanese beetle belongs to the family Scarabaeidae in the order Coleoptera. The beetle species *Popillia japonica* is commonly known as the Japanese beetle. Japanese Beetle, insect pest of the scarab family. The Japanese beetle was introduced

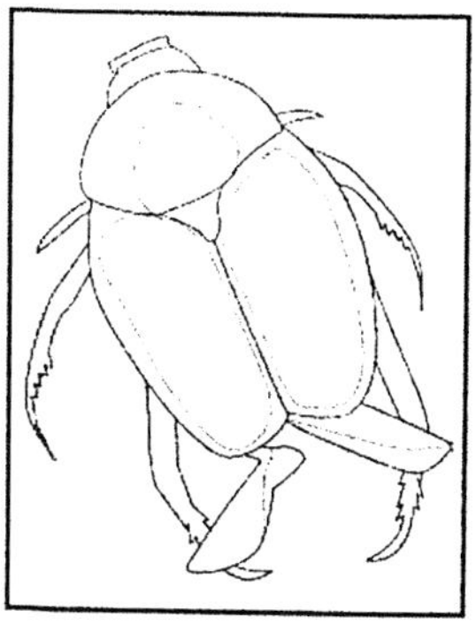

accidentally into New Jersey from Japan in 1916 and is now widespread over the eastern United States. The adult has a broad, thick body about 1.3 cm (about 0.5 in) long and is iridescent green, with tan wing covers. It attacks the foliage of a wide variety of plants and also eats their fruit, causing widespread destruction. The Japanese beetle is most common in July and August, at which time the female lays its eggs 13 to 15 cm (5 to 6 in) below the surface of the ground. The white grubs (wormlike larvae) feed on the roots of grasses until fall, when they descend 30 cm (12 in) or more underground and hibernate; they emerge in midspring and pupate in June after attaining a length of slightly less than 2.5 cm (less than 1 in). It is a clumsy flier, dropping several centimcters when it hits a wall. Parasitic wasps, largely responsible for the control of the insect in Japan, were imported into the United States in the 1920s; they reduce the beetle population by preying on the grubs. The grubs are also subject to certain diseases caused by bacteria, fungi, nematodes, and viruses. The most important biological controls are milky disease and a virus infection called blue disease.

Japanese beetle traps therefore consist of a pair of crossed walls with a bag underneath, and are baited with floral scent, pheromone, or both. However, studies done at the University of Kentucky suggest that traps attract more beetles than they actually trap, thus causing more damage along the flight path of the beetles and in the vicinity of the trap than may have occurred if the trap was not present. These insects damage plants by skeletonizing the foliage, that is, consuming only the leaf material between the veins.

JUNE BUG

June bugs belong to the family Scarabaeidae of the order Coleoptera. May bugs are classified in the genus Phyllophaga.

The green June beetle is classified as *Cotinis nitida*. June Bug, common name for any of several beetles in the scarab family, also called June beetle. The adults are most common in June. In the northern United States, the name is applied to the numerous species that are known as May bugs in the southern United States, where they emerge earlier. The brown, stout-bodied adults are about 25 mm (about 1 in) long and feed on leaves. The larvae, known to horticulturists as white grubs, burrow in soil, feed on the roots of plants, and often damage grass lawns. The larval stage persists for two to three years.

Adult beetles, commonly referred to as May beetles or June bugs are ½ to 5/8 inches long, and reddish brown. White grubs are "C"-shaped larvae, up to 1 inch long, with cream-coloured bodies and brown head capsules. They have three pairs of legs, one on each of the first three segments behind the head. There are more than 100 species of scarab beetles from several genera (*e.g.,Cyclocephala, Phyllophaga* and others) in Texas that are considered to be white grubs, May beetles and June bugs. However, the most common is *Phyllophaga crinita*. Their biologies are similar, but species differ in distribution, habitat preference, length of life cycle and seasonal occurrence. Other common species include the **southern masked chafer**, *Cyclocephala immaculata* (Oliver), and the **green June beetle**, *Cotinis nitida* (Linnaeus). The **Japanese beetle**, *Popillia japonica*, introduced into the northeastern United States and migrating west and south, has recently been detected in some Texas counties.

Adults begin to emerge in spring. During adult flights large numbers of beetles can be attracted to lights. Peak flights occur in mid to late June in central Texas. Females, less attracted to lights, tunnel 2 to 5 inches into the soil and deposit eggs. In 3 to 4 weeks, small grubs (larvae) hatch from eggs and develop through three stages (instars), with the first two stages lasting about 3 weeks. The last larval stage remains in the soil from the fall through spring. In spring and early summer, white grubs pupate 3 to 6 inches deep in the soil. Adults emerge from pupae in about 3 weeks. There is one generation per year, but in north Texas, development may take two years.

Habitat, Food Source(s), Damage

Mouthparts are for chewing. *Phyllophaga crinita* is common in Texas turfgrass, particularly Bermudagrass, St. Augustine grass and tall fescue. Feeding of large numbers of grubs causes lawns to turn yellow and die. Severely damaged grass can be "rolled up" like a carpet. Grubs also feed on the roots of weeds, vegetable transplants and ornamental plants. In agriculture, they are important pests of forage, corn, sorghum and sugarcane. Most severe injury to plants is caused by large (third stage or instar) grubs feeding on roots in the fall and spring. White grubs are frequently encountered tilling garden soil or by sifting through soil underneath damaged turfgrass. Adults can be abundant around lights in the spring of the year.

LADYBIRD BEETLE

Ladybird beetle (family Coccinellidae), also called ladybug, any of approximately 5,000 widely distributed species of beetles (insect order Coleoptera) whose name originated in the Middle Ages, when the beetle was dedicated to the Virgin Mary and called "beetle of Our Lady." Ladybird beetles are hemispheric in shape and usually 8 to 10 mm (0.3 to 0.4 inch) long. They have short legs and are usually brightly coloured with black, yellow, or reddish markings. The colour of the wing covers and number of spots vary among species. The pattern of the nine-spotted ladybird beetle (*Coccinella novemnotata*), which has four black spots

Ladybird Beetle or Ladybug, common name for any of numerous related species of brightly coloured beetles found in temperate and tropical regions throughout the world. The ladybird beetle is less than 1.2 cm (less than 0.5 in) in maximum length. It has a nearly hemispherical body, rounded above and flat below, a small head, and short legs. Ladybird beetles are often red or orange above, spotted with black, white, or yellow. Some species are black, with or without spots. The larvae are also brilliantly coloured, often blue, with stripes of orange or black.

All the ladybird beetles, with the exception of the members of one vegetation-eating genus, are carnivorous. In both the adult and larval stages they feed on insects harmful to plants, such as aphids and scale insects. Because of the help ladybird beetles render farmers in destroying agricultural pests, the beetles were popularly regarded in the Middle Ages as instruments of benevolent intervention by the Virgin Mary, whence the common name ladybird.

A common North American species of ladybird beetle, the nine-spotted ladybug beetle, is orange above, spotted with black. Adults of the two-spotted ladybug beetle often hibernate in houses during winter. This beetle is orange above, with a single large black spot on each elytron (protective outer wing). The convergent ladybug beetle is a western American species, the adults of which commonly swarm in large numbers on mountain peaks. These swarms are collected by western agricultural firms and are distributed to farmers for aphid control. The vedalia, an Australian species, has been imported into California to fight the cottony-cushion scale insect, which attacks citrus trees.

LEAF BEETLE

Beetles in the family **Chrysomelidae** are commonly known as **leaf beetles**. This is a family of over 35,000 species in more than 2,500 genera, one of the largest and most commonly-encountered of all beetle families. Leaf beetles are partially recognizable by their tarsal formula, which appears to be 4-4-4, but is actually 5-5-5. They are distinguished with difficulty from long-horned beetles (family Cerambycidae) by the antennae not arising from frontal tubercles. Adult and larval

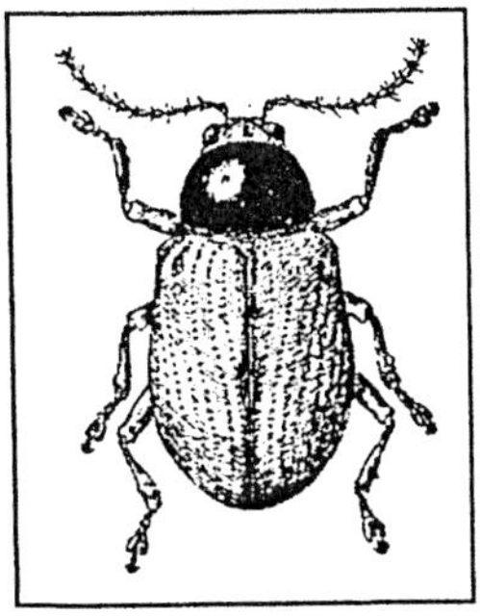

leaf beetles feed on all sorts of plant tissue. Many are economically important pests of agriculture, for example the Colorado potato beetle (Leptinotarsa decemlineata), the asparagus beetle (Crioceris asparagi), the cereal leaf beetle (Oulema melanopus), and various flea beetles.Leaf beetles eat only vegetation, especially leaves; some are agricultural pests. They usually have small, plump bodies, with smooth, metallic surfaces, and short legs. The insects lay their eggs on the ground, and the larvae usually pupate on or in the ground. Some species are aquatic. About 1400 species of leaf beetles are found in North America.

LONGHORNED BEETLE

The longhorn beetles (Cerambycidae; also known as long-horned beetles or longicorns) are a cosmopolitan family of beetles, typically characterized by extremely long antennae, which are often as long as or longer than the beetle's body. Longhorned beetles are in the family Cerambycidae, order Coleoptera. The roundheaded appletree borer is *Saperda candida*; the red milkweed beetle is *Tetraopes tetrophthalmus*; the California prionus is *Prionus californicus*; the valley elderberry longhorned beetle is *Desmocerus californicus*. In various members of the family, however, the antennae are quite short (e.g., *Neandra brunnea*, figured below) and such species can be difficult to distinguish from related beetle families such as Chrysomelidae. The family is large, with over 20,000 species described, slightly more than half from the Eastern Hemisphere. Several are serious pests, with the larvae boring into wood, where they can cause extensive damage to either living trees or untreated lumber (or, occasionally, to wood in buildings; the old-house borer, *Hylotrupes bajulus*, being a particular problem indoors).

A number of species mimic ants, bees, and wasps, though a majority of species are cryptically coloured. The rare giant long-horned beetle (*Titanus giganteus*) from northeastern South America is often considered the largest (though not the heaviest, and not the longest including legs) insect, with a maximum known body length of just over 16 centimeters. Longhorned Beetle, common name for any member of a family of beetles known for their long antennae, or "horns," and the wood-boring habits of the larvae. The larvae are known as roundheaded wood borers. Longhorned beetles live throughout the world.

There are about 20,000 species, more than 1200 of which occur in North America. Adult longhorned beetles vary greatly in shape, size, and color. In general they are cylindrical and rather long-bodied. Some tropical species are among the world's largest beetles, reaching 15 cm (6 in) or more in length. But most species measure 8 to 50 mm (0.33 to 2.00 in) long. Some species of lorghorned beetles are dull brown, while others have bright colours or intricate patterns on their wing covers. The beetle's prominent antennae are at least two-thirds as long as the body, and in some species the antennae are longer than the body. Males have longer antennae than females. Some species of longhorned beetles are slow-moving, while others can run quickly or are strong fliers.

Depending on the species, adults may be observed feeding on flower parts, leaves, or bark. Females lay eggs in bark crevices or in soil near roots. Newly hatched larvae immediately begin to excavate into the plant they feed on. Some species feed only on dead or decaying wood, while others bore into roots or beneath bark of living trees and shrubs. Larvae are pale-coloured, often creamy white, and wormlike. The larvae of large species can grow to several inches in length. Their small round head and cylindrical form distinguishes them from flatheaded borers (larvae of metallic wood borers), which have an enlarged and flattened thorax that is wider than the rest of the body. In some countries, roundheaded borers are eaten by humans as a delicacy.

Like all beetles, longhorned beetles have complete metamorphosis. Larvae commonly pupate in a chamber they construct under bark. Most species require one to three years to complete one generation, although some can develop from egg to adult in as little as two or three months. Mature larvae of a few species can remain inactive in wood for many years. There are records of longhorned beetles emerging from wood beams or furniture in houses after remaining dormant for more than two decades.

Some species of longhorned beetles are pests of forest, shade, or fruit trees and certain shrubs. The roundheaded appletree borer, for example, infests apple trees. Protecting trees from injury and otherwise providing plants with proper care usually will prevent damage by longhorned beetles. Most longhorned beetles attack only injured or weakened woody plants. Longhorned beetles help to decompose dead and dying trees, thereby making nutrients and sunlight available for new plant growth.

Longhorned beetles have many natural enemies, especially parasitic wasps and the larvae of certain other beetles. Many birds feed on adults and woodpeckers are fond of the larvae. Lizards sometimes lie in wait and capture adults when the beetles land on bark to mate or lay eggs.

The brilliantly coloured red milkweed beetle is frequently seen on milkweed plants in North America. The *California prionus* is the largest North American beetle, reaching 6 cm (2.3 in) in length. The huge adults have a spiny neck and can fly. Their larvae bore in the roots and stumps of many species of hardwood and conifer trees.

Brightly coloured adults of many species of longhorned beetles are a favorite among insect collectors, probably exceeded in popularity only by butterflies and large moths. Partly because of their relatively slow rate of reproduction, the survival of some species of longhorned beetles is endangered. For example, the valley elderberry longhorned beetle is threatened with extinction in California. Development and farming along riverbanks have destroyed much of this beetle's natural habitat.

METALLIC WOOD BORER

Metallic wood-borers are so-called because adults of many species have a metallic luster, giving them a shiny appearance while at rest. The name wood-borer refers to the larval stage. Larvae bore under bark or deeper into heartwood of a variety of tree species. The term "flat-headed wood-borers" is also used with these beetles, and refers to the dorsally flattened appearance of the larvae. Adults are quite active and quick to take flight when disturbed. Activity generally occurs during the hottest part of the day, and individuals may be seen sunning on logs or stumps. Many species also occur on plants, and may be collected by beating foliage. Members of the genus *Chalcophora* are the largest buprestids in Florida. These species (20-30mm) of metallic woodborers are frequently encountered by collectors in Florida. Individuals may be seen resting on sidewalks, on walls, and may actually land on people, especially if they walk around in bright clothing. This may be quite disconcerting to those that are intimidated by large insects. Florida species are attracted to turpentine-baited traps. Metallic wood borers comprise the family Buprestidae, order Coleoptera. The flatheaded appletree borer is *Chrysobothris femorata*; the Pacific flatheaded borer is *Chrysobothris mali*. The bronze borers are in the genus Agrilus; and the twolined chestnut borer is *Agrilus bilineatus*.

Biology

Members of Buprestidae utilize a variety of host plants as food and as larval habitat. *Chalcophora*, both *georgiana* and *virginiensis* are reported as general pests of a variety of species of Pine. Blatchley (1910) stated that *C. liberta* frequented scrub pine (*Pinus banksiana* Lamb.) in the dune region of Lake, Porter, and Laporte counties, Indiana. Adults fed on young buds and leaves of the pine while larvae lived as borers in the decaying wood. Females oviposit in cracks and crevices in the bark or in wounds. Larvae develop under bark, then bore deeper into heartwood where they pupate in a single cell. Adults emerge from the host tree leaving a characteristically shaped elliptical exit hole.

The genus *Chalcophora* Dejean 1833 contains five species; *Chalcophora fortis*, *C. angulicollis*, *C. georgiana*, *C. liberta*, and *C. virginiensis*. The first two are northern and/or western species that do not occur in Florida. The remaining three species have been recorded from Florida. There is some question concerning the occurrence of *C. liberta* in Florida. Literature records exist for this species but no recent verifiable specimens have been seen. The checklist of the Chalcophorina of North America (north of Mexico) lists *C. liberta* and *C. virginiensis* in Florida, with *C. georgiana* known from Georgia and Mississippi.

ROSE CHAFER

Rose chafer - a common metallic green European beetle: larvae feed on plant roots and adults on leaves and flowers of roses. The rose chafer, *Macrodactylus subspinosus*, is commonly found in many areas of Minnesota, particularly areas with sandy soil. It is a pest on many different types of flowers, fruits, trees and shrubs. An adult rose chafer is a moderate-sized insect, measuring between 5/16-inch to almost 1/2-inch in length. It's a slender beetle, pale green to tan in color with reddish brown or orangish spiny legs. It has short, lamellate antennae, *i.e.* a series of flat plate- or page-like segments. A rose chafer sometimes resembles a wasp when it's flying. The larval stage is called a grub and has a brown head and conspicuous legs. Like other grubs, it's body is bent into a 'C' shape. Fully grown, a rose chafer larva is about 3/4-inch long. Rose chafer larvae are rarely, if ever, seen.

Adults emerge from the ground in late May and early June. Because the rose chafer prefers sandy soil to lay eggs, plants located on sandy sites are most likely to be attacked. Adult beetles feed on plants for three or four weeks, generally until late June. Females lay eggs in the soil, then die shortly afterwards. Two to three weeks later, the eggs hatch into small, white grub like larvae which feed on the roots of grasses and weeds. The larvae spend the winter in the soil below the frost line. They pupate the following spring and then emerge as adults. There is one generation a year.

SCARAB

Scarabs make up the family Scarabaeidae, in the order Coleoptera. Most June beetles are classified in the genus Phyllophaga. The scarab sacred in ancient Egypt is classified as Scarabaeus sacer. Scarab, common name for any of various large, bright-coloured beetles of which about 30,000 are found worldwide. In addition to their brilliant coloring and bulky size, up to 15 cm (6 in) in length, scarabs are distinctive for diverse horns or protuberances growing on the head or thorax; those on the males of many species are used in combat. Scarabs are divided into two groups, the scavenging dung beetles, which feed on decaying plants, and the chafers, which feed on young plants.

Dung beetles are beetles that feed partly or exclusively on feces. All of these species belong to the superfamily Scarabaeoidea; most of them to the subfamilies Scarabaeinae and Aphodiinae of the family Scarabaeidae. This beetle can also be referred to as the scarab beetle. As most species of Scarabaeinae feed exclusively on feces, that subfamily is often dubbed true dung beetles. There are dung-feeding beetles which belong to other families, such as the Geotrupidae (the earth-boring dung beetle). The Scarabaeinae alone comprises more than 5,000 species.

In Egypt, an important species of scarab was held sacred by the ancient, sun-worshiping Egyptians, who regarded the beetle as a symbol of immortality. Amulets, seals, and rings in the shape of the beetle came to be known as scarabs.

Many dung beetles, known as *rollers*, are noted for rolling dung into spherical balls, which are used as a food source or brooding chambers. Other dung beetles, known as *tunnelers*, bury the dung wherever they find it. A third group, the *dwellers*, neither roll nor burrow: they simply live in manure. They are often attracted by the dung burrowing owls collect.

STAG BEETLE

Stag Beetle, common name for beetles having long, antlerlike mandibles or jaws, used in combat. Stag beetles make up the family Lucanidae of the order Coleoptera. The reddish-brown stag beetle as *Pseudolucanus capreolus*, the giant stag beetle is classified as *Lucanus elephus*, the black stag beetle as *Pseudolucanus placidus*, and the giraffe stag beetle as *Cladognathus giraffa*. These mandibles are sometimes as long as the body of the insect, which in some tropical species is as much as 10 cm (4 in). The females have smaller jaws, which they use to dig holes in rotting wood for the deposit of eggs. Stag beetles are usually black, brown, bright red, green, or yellow in color, and despite their cumbersome jaws, they are able to fly. About 900 species are found throughout the world, 30 of them in North America. Stag beetles thrive especially in dank forests where many decaying logs are found. Among the better known forms in the United States are three species of giant stag beetles: the giant stag beetle, the reddish-brown stag beetle, and the black stag beetle. The giraffe stag beetle is found in India and Java.

Male stag beetles use their jaws to wrestle each other for favoured mating sites in a manner that parallels the way stags fight over females. Fights may also be over food, such as tree sap and decaying fruits. Despite their often fearsome appearance they are not normally aggressive to humans.

Female stag beetles are usually smaller than the males, with smaller mandibles. As larvae, females can be distinguished from males by the presence of cream-coloured, fat ovaries visible through the skin around 2/3 down the larva's back.

The larvae feed for several years on rotting deciduous wood, growing through 3 larval stages until eventually pupating inside a pupal cell constructed from surrounding wood pieces and soil particles. In the final larval stage, "L3", the grubs of larger species, such as *Prosopocoilus giraffa*, may be the size of a human finger

TIGER BEETLE

Tiger beetles make up the family Cincindelidae of the order Coleoptera. The largest American tiger beetle is classified as *Amblycheila cylindriformis*. More than half of all tiger beetles are classified in the genus Cicindela. The tiger beetles are a large group of beetles known for their aggressive predatory habits and running speed. The fastest species of tiger beetle can run at a speed of 9 km/h (5.6 mph), which, relative to its body length, is about 22 times the speed of former Olympic sprinter Michael Johnson. Tiger Beetle, common name for predacious beetles that are generally metallic golden-green in color, with spots or stripes on the wing covers. About 1500 species are known, the majority inhabiting the tropics. Some species are wingless, whereas others are active fliers. Most are generally found in sandy terrain, some only on the mounds of termites, and others on the trunks of trees, which they ascend in a spiral manner. Both adults and larvae prey on other insects, and adults also eat worms and snails.

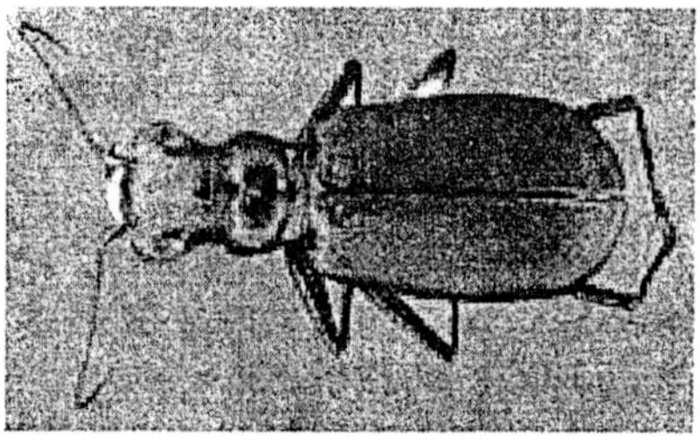

The larvae of many dig tubes in moist sand, closing the holes with their heads, and seize unwary passing insects and drag them below to be devoured. The largest American tiger beetle is found in sandy regions in the midwestern United States. One genus contains more than half of the species in the entire family and includes most of the forms that are found in North America.

Tiger beetles were traditionally classified as the family Cicindelidae but most authorities now treat them as the subfamily Cicindelinae of the Carabidae (ground beetles). The most recent classifications, however, have relegated them to a monophyletic subgroup within the subfamily Carabinae, though this is not yet universally accepted; accordingly, there is no consensus classification for this group, at any level from family down to subspecies, and it can be exceedingly difficult to decipher the taxonomic literature surrounding this group. In mid 2005, There was about 2,600 species known. The richest biogeographical region is the Oriental (Indo-Malayan) region, with 753 species, followed by the Neotropical Region (495), the Ethiopian region (459), the Australo-Papouan and Wallacea Region (353), Madagascar and Mascarene Is.(217), the Neartic Region (182) and the Paleartic Region (131). The discovery of many new species during the past decade increased the number of known species. For the Ethiopian Region only, from 396 species in the year 2000 to 477 species in 2010.

WATER BEETLE

Water Beetle, common name for any of numerous aquatic beetles including the true, or diving, water beetles, the whirligig beetles, and the water scavenger beetles. A water beetle is a beetle adapted to living in water. Water beetles rise to the water surface and take atmospheric air into their tracheal systems. There are approximately 2000 species of water beetles. The rest marine species tend to live in the intertidal zone. Water beetles are black, brown, or greenish and .08 to 1.57 inches long. Many water beetles carry an air bubble underneath their abdomens, which both provides an air supply, and prevents water from getting into the spiracles. Others have the surface of their exoskeleton modified to form a plastron, or "physical gill", which permits direct gas exchange with the water. The hind legs are strong like the grasshoppers except it is not for jumping but for swimming. The wing covers are usually grooved on the

female and smooth on the male. They have fringed hind legs with hair adapted for swimming.

Adults hibernate in underwater soil during a good portion of the year. Water beetles are attracted to lights at night. Most families of water beetles have larvae that are also aquatic; many have aquatic larvae and terrestrial adults.

Whirligig beetles are native to the temperate and tropical regions throughout the world, spending most of their life at the surface of the water. Their eyes are divided for vision in both water and air, and they can dive to considerable depths. Like the diving beetles, they are equipped with an air cavity under the wing covers. They hibernate during the cold months of the year, but emerge periodically for their characteristic whirling dances. The adults are lustrous black above and are yellow, black, or brown below. The body, like that of the diving beetles, is elliptical and convex.

Water beetles belong to the order Coleoptera. True, or diving beetles make up the family Dytiscidae. Types in which the adults are not necessarily aquatic include Hydrophilidae, Lutrochidae, Dryopidae, Elmidae, Eulichadidae, Heteroceridae, Limnichidae, Psephenidae, Ptilodactylidae and Sphaeriusidae. A nickname that people call the water beetle is the water devil. Also because of there rapidity they are also known as water tigers.

WEEVIL

Weevils belong to the superfamily Curculionoidea of the order Coleoptera. The largest family is Curculionoidae. The pea weevil is classified as *Bruchus pisorum;* the bean weevil as *Acanthoscelides obtectus;* the tiny granary weevil is classified as *Sitophilus granarius,* the rice weevil as *Sitophilus oryzae,* and the alfalfa weevil as *Hypera postica.* A weevil is any beetle from the Curculionoidea superfamily. They are usually small, less than 6 millimetres (0.24 in), and herbivorous. There are over 60,000 species in several families, mostly in the family Curculionidae (the true weevils). Some other beetles, although not closely

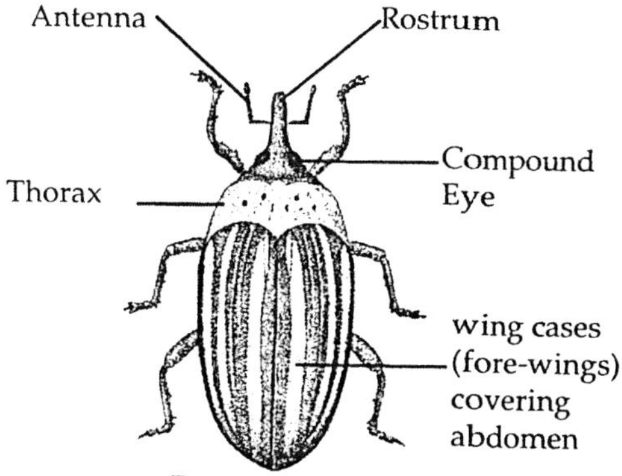

Bean stem weevil

related, bear the name "weevil", such as the biscuit weevil (*Stegobium paniceum*), which belongs to the family Anobiidae. Many weevils are damaging to crops. The grain or wheat weevil (*Sitophilus granarius*) damages stored grain. The boll weevil (*Anthonomus grandis*) attacks cotton crops. It lays its eggs inside cotton bolls, and the young weevils eat their way out.

Weevils are often found in dry foods including nuts and seeds, cereal and grain products, such as pancake mix. In the domestic setting, they are most likely to be observed when a bag of flour is opened. Their presence is often indicated by the granules of the infested item sticking together in strings, as if caught in a cobweb.

Types of Weevils

Granary Weevils

This is a common type of weevil. It comes in black or dark brown. At about 4 mm, it is noticeably smaller compared to other members of the species. Granary weevils, or grain weevils, are known all over the world because they infest most types of grains. In India, they are notorious for decimating rice harvests, so they are also referred to as rice weevils in that country. A female granary weevil lays its eggs inside grains and can have an average of six reproductive cycles every year. At home, you can find this type of weevil in your kitchen eating their way into cereal boxes or cracker packs.

Pea Weevils

Members of this group are closer to being regular leaf beetles than weevils. But just like other weevil varieties, the female pea weevil would bore deep into seeds, in this case peas, to deposit its eggs. They are known to feed on peas and even beans.

Apple Blossom Weevils

This weevil originated from Europe and some parts of Asia and North Africa. It took its name from the fact that it feeds on apple or even pear blossoms. Adult apple blossom weevils are normally 4 mm in length with long and narrow snouts. They come in dark red or brown color with white markings on both wings. They can get inside your home if the apples or pears you buy from the grocery are not thoroughly cleaned or if you live near an apple or pear plantation. The flowers that have been attacked by this weevil will fail to develop, thus bringing down the number of fruits that can be harvested. On top of using apple or pear flowers as food, the females of this type of weevil also deposit their eggs inside.

Leaf Weevils

These weevils are characterized by short snouts and metallic green or bronze scales, but at first glance they appear to be black or dark brown in color. They can grow anywhere from 4 to 9 mm and are common in gardens. As their name suggests, these weevils eat the leaves of trees like apple, pear, oak, alder, cherries and even rhododendrons. As they feed, they can also cause damage to blossoms. The good news is that leaf weevils can be easily controlled using garden pesticides.

Crop Damage

The bean stem weevil (*Alcidodes leucogrammus*), bores into the base of the stem of French bean (Phaseolus) or cowpea

(Vigna) plants and lays a small number of eggs. These hatch in about 6 days to larvae. The larvae feed on the stem tissue and damage the plant, sometimes causing the stem to break off or fall over. They pupate in the stem and emerge as adults in about 50 days, though this period is very variable. This weevil may attack and damage several rows of garden beans but it is not usually a serious agricultural pest. Burning infected plants at once, and all plants at the end of the growing season, will help to control the weevil. *Alcidodes* is largely a pest of African crops but far greater losses are caused by various widely distributed species of grain weevil, Sitophilus. These are smaller and less conspicuous than *Alcidodes* but they do more damage because they lay their eggs in the fruits or seeds of beans, maize or rice as the crops are growing in the field. While the seeds or grains are in store, the larvae eat the inside of the seed.

In the seed they pupate and emerge as adults in as little as 10 days. The adults then bore into the stored grain all around them and lay their eggs. Thus, in a month or two a large proportion of the grain can be damaged or destroyed. This is one reason why old seed and new seed should not be mixed; any weevils hatching from the old grains will attack the new seed.

Control

Weevils cause millions of dollars worth of damage every year. Common methods of control are burning infested fruits, nuts, or stems; plowing the ground in which the insects pupate; and spraying crops with insecticides.

Booklouse

Kingdom: Animalia, **Phylum**: Arthropoda, **Class**: Insecta, **Superorder**: Psocodea, **Order**: Psocoptera

Booklice make up the order Psocoptera, which is divided into several families. The cereal psocid belongs to the family Liposcelidae and is classified as *Liposcelis divinatorius*.

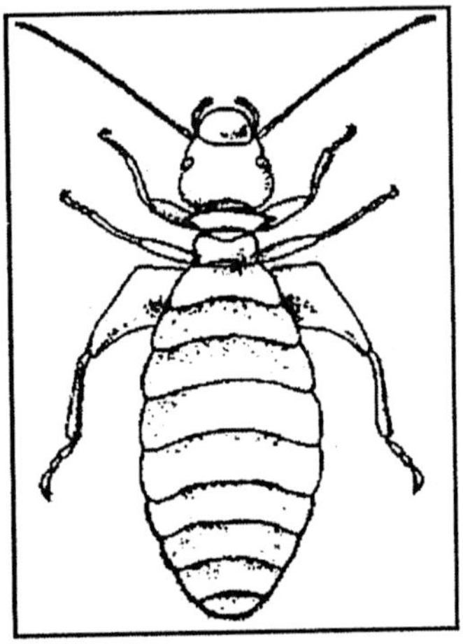

Booklouse, common name for any of an order of small, delicate insects known for feeding on dead plant and animal matter or starchy substances such as bookbinding paste. Booklice are also called barklice or psocids. The Latin name of their order, Psocoptera, means "small rub" (psoco) and "winged" (-ptera), terms that refer to the abrasionlike damage caused by their feeding and the fact that some have wings. They vaguely resemble true lice but do not feed on other live animals. Booklice are found throughout the world but are most diverse in tropical regions. About 2600 species are known worldwide, and 245 species are found in North America.

Psocids or Booklice have long, filamentous antennae and a characteristic bulging clypeus (the area just above the mouthparts). They have chewing mouthparts and the wings of domestic species are usually absent. They range in size from

1/25 to 1/13 of an inch (1 to 2 mm) in length. Coloration is from almost colorless to gray or light brown.

The adults are small 1 to 2 mm cream coloured, wingless insect readily seen moving over paper, and have a characteristic square shaped head. Booklice generally feed on small fungal growths that develop on a wide variety of materials, particularly if they are damp, including glues, paper, books, woodwork, pressed pine, leather and plaster. They even feed on small cellulose particles. There is also a species that will breed in large number on stored products such as cereals, flour, oats and brans. Book lice are a nuisance without causing any damage . They are thought to bite humans, but this has been difficult to prove.

Food

Psocids feed upon microscopic molds. Thus, any manufactured material of plant origin that would support the growth of these molds is susceptible to their attack. They are found in nature on the bark of trees and shrubs, preferring damp, warm, undisturbed environments. They can commonly be found in books and book bindings, storage boxes, pape: and goods

Life Cycle

The adult lays her eggs singly. They are covered with a sticky substance and cling to the food surface. On hatching, the nymphs, which appear as miniature adults. After a number of moults, up to five, the mature adult stage is reached. Ideally, the life cycle may be completed in 3 to 4 weeks. Because of the short reproductive period, high populations may be reached quickly if they are left undisturbed. Booklice are soft-bodied insects less than 6 mm (0.25 in) long; most are minute, only 1 to 2 mm (0.4 to 0.8 in) in length. They are usually gray or yellowish. Adults have wings in some species and are wingless in others. In some species, both adult forms occur. Psocids are sometimes confused with similar-looking psyllids. However, psocids have chewing mouthparts and commonly run or fly away when disturbed; psyllids, like aphids, suck plant juices and jump when alarmed. When viewed through a hand lens, booklice can be recognized by the distinctly swollen, bulbous area at the front of their face between the two widely spaced antennae.

Females lay pale, oval eggs singly or in clusters, sometimes covered with silk webbing or bits of soil or leaves. Booklice have incomplete metamorphosis. Most species develop through six nymphal instars, or growth stages, as they gradually change from egg to nymph (an immature form with some adult characteristics) and then to adult. Some species live in colonies under protective silken webs. These webs resemble sheetlike spiderwebs and may cover rough tree bark. Booklice also live on leaves, rocks, and in soil. Entomologists collect psocids by sifting soil debris, by sweeping trees and shrubs with a net, or by shaking branches to dislodge them onto a sheet of white paper or other collecting surface held underneath the plant.

Most booklice are of relatively little economic importance, as they feed only on pollen; various fungi, such as sooty mold; and decaying plant or animal tissue. The cereal psocid is an occasional pest. It feeds on cereal, wallpaper, bookbindings, and collections of dead insects, especially if these materials are slightly damp. The cereal psocid looks like a minute brown speck, less than 1 mm (0.04 in) long.

Habitat

Psocids are found on all continents, including Antarctica. There are 4,408 species of psocids worldwide, mostly in the tropics. About 260 species occur in the United States and Canada. Psocids live in a wide variety of habitats on land. In spite of the common names that include the word "louse," these insects do not live on other animals. They are most common on dead or living leaves, on stone or bark surfaces, and in leaf litter.

Some species prefer living in caves; others are known to bore into wood. A few species of psocids are common in homes and buildings, especially where food is stored. Psocids eat lichen or mosses, funguses, and other bits of plant and animal tissues. They will even feed on the skin flakes of humans and their animals. Some species scavenge dead insects or eat their eggs. Most psocids live in the wild and are seldom, if ever, noticed by humans. However, species common in homes are often considered household pests. They can reproduce rapidly under warm, humid conditions and will become serious pests in foods stored in cupboards and pantries.

These species are known to cause sneezing, coughing, itching, and rashes, as well as asthma attacks in sensitive people. No species of psocids is endangered or threatened. Still, many species, especially those living only on single islands or in caves, could easily become threatened by habitat destruction due to bad weather or human activities.

Control

Ventilation and drying largely control book lice as most populations are usually associated with high moisture levels. If this fails spraying the back of cupboards and shelving with an insecticide will help keep the book lice numbers down. In some cases where book lice infestation is too large to treat, fumigation is the only method of treatment left. This should only be used as a last resort.

Bookworm

Family: Bostrichidae, **Order**: Coleoptera.

Bookworm, any insect (*e.g.*, moths, beetles) whose larval (or adult) forms injure books by gnawing the binding and piercing the pages with small holes. No single species may properly be called the bookworm because a large number of insects feed upon dry, starchy material or paper and may damage books. Among the most widely known bookworms are the silverfish (order Thysanura) and the booklice (order Psocoptera). Termites and roaches are also frequently guilty. The beetle larvae known as bookworms belong to the family Bostrichidae, of the order Coleoptera. The moth larvae known as bookworms belong to the family Oecophoridae, of the order Lepidoptera. Booklice make up the family Liposcelidae, of the order Psocoptera.

Bookworm is a popular generalization for any insect which supposedly bores through books. Actual book-borers are uncommon. Both the larvae of the death watch beetle (*Xestobium rufovillosum*) and the common furniture beetle (*Anobium punctatum*) will tunnel through wood and paper if it is nearby the wood. A major book-feeding insect is the book or

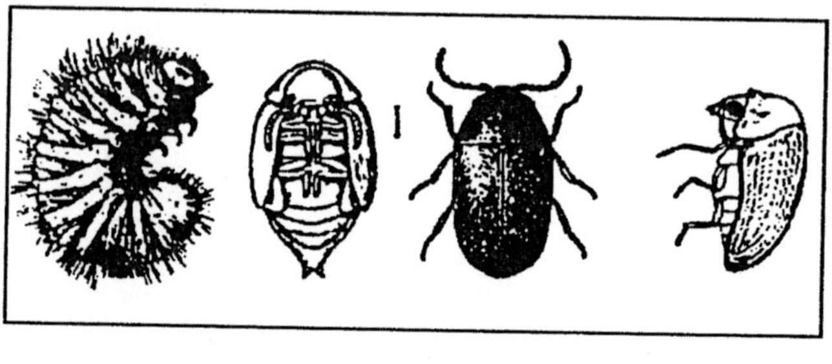

paper louse (aka booklouse or paperlouse). A tiny (under 1 mm), soft-bodied wingless *Psocoptera* (usually *Trogium pulsatorium*), that actually feeds on microscopic molds and other organic matter found in ill-maintained works (e.g., cool, damp, dark, and undisturbed areas of archives, libraries, and museums), although they will also attack bindings and other book parts. It is not actually a true louse. Many other insects, like the silverfish (*Lepisma saccharina*) or cockroach (various *Blattodea*), will consume these molds and also degraded paper or the starch-based binding pastes – warmth and moisture or high.humidity are prerequisites, so damage is more common in the tropics. Modern glues and paper are less attractive to insects. Two moths, *Tineola bisselliella* and *Hofmannophila pseudospretella*, will attack cloth bindings. Leather-bound books attract various other consumers, such as *Dermestes lardarius* and the larvae of *Attagenus unicolor* and *Stegobium paniceum*. The bookworm moth (*Heliothis zea* or *H. virescens*) and its larvae are not interested in books. The larvae are pests for cotton or tobacco growers as the cotton bollworm or tobacco budworm. The beetle larvae are about 0.25 to 0.50 cm (about 0.1 to 0.2 in) in length. Insects other than beetles, such as cockroaches and the tiny larvae of certain moths, have also been known to destroy books. One family of tiny, soft-bodied, wingless insects in particular are known as booklice and do considerable damage when they feed on book bindings or on biological specimens found in museums.

Eggs are deposited near the surface of the binding or on the edges of the leaves and hatch in 5 or 6 days. The young grubs penetrate the book and tunnel their way up the spine or on the inside of the cover. After becoming full grown and before transforming to pupae, the grubs make their way close to the exterior surface. A pupal chamber is then hollowed out by the grub, and here the grub transforms into a pupa. The adult beetle emerges from the pupal chamber by chewing a small hole through the exterior of the binding. Approximately 2 months are needed for a complete life cycle from egg to adult. Four or more generations a year are not unusual, particularly in the tropics.

Bug

Bug (insect), name properly applied to any member of a particular order of insects, including bed bugs, lace bugs, red bugs, squash bugs, chinch bugs, and many so-called water bugs. These are the true bugs, although the name bug is sometimes loosely used to refer to their close relatives-members of the order including the cicadas, aphids, tree hoppers, lantern flies, mealybugs, white flies, and scale insects. The name bug is also often incorrectly applied to beetles (for example, ladybug and potatobug), to other insects (doodlebug), and even to animals that are not insects (sowbug and pillbug). Technically, or taxonomically, a bug is an insect belonging to the order Hemiptera, known commonly as the true bugs. Aphids, cicadas, assassin bugs, and a variety of other insects claim membership in the order Hemiptera. All bugs are insects, but not all insects are bugs. Beetles are not true bugs, nor are butterflies or bees or flies. The word bug is widely used to describe just about any arthropod, from decidedly non-insect critters like millipedes and pillbugs to arachnids like spiders and ticks.

In the true bugs the basal half of the wing is usually thickened like the wing covers of beetles, although the tips of the wings are membranous-that is, they consist of a thin layer of tissue. In the order of aphids and cicadas the entire wing is membranous. Both orders are characterized by sucking mouthparts in the form of well-developed beaks. Although a few, such as bed bugs, are parasitic on vertebrates, most bugs subsist on plant juices, and many are among the most economically harmful of insects. A few bugs, such as the assassin bug, are predatory and are therefore useful in the biological control of pests. In some tropical countries certain bugs, such as the cochineal insects, are cultivated for the valuable dye they secrete. The metamorphosis of most bugs is an incomplete process-that is, the larvae develop gradually into adults without any intervening pupal stage.

Life Cycle

The methods of reproduction of some bugs are unusual and complicated. The eggs of some aphid species, for example, hatch during the spring into wingless females; these females reproduce by parthogenesis-that is, without males. After several such generations within one season on one kind of plant, the eggs hatch into winged females that then fly to a different plant and reproduce more wingless females. In the latter part of the season more winged females are produced, and they migrate back to the original species of host plant. Only late in the season are males and females produced that have the ability to reproduce sexually. The eggs from the mating of these insects lie dormant during the winter months and hatch in the following spring to begin a new cycle.

Some bugs are very large; certain aquatic species reach a length of nearly 13 cm (nearly 5 in). Some bugs secrete a substance characteristically pungent, penetrating, and often having a disagreeable smell. Members of the squash-bug, stinkbug, and shield-bug families have a particularly bad odour. Certain plant-sucking bugs cover themselves in their immature stages with a mass of foam known as frog spittle. Some bugs secrete a sweet-tasting fluid called honeydew, and these bugs are often herded, protected, and "milked" for the honeydew by certain species of ants. Some bugs exhibit the unusual characteristic among insects of caring for their young.

ASSASSIN BUG

Common Name: Assassin Bug, **Order Name**: Hemiptera **Family Name**: Reduviidae.

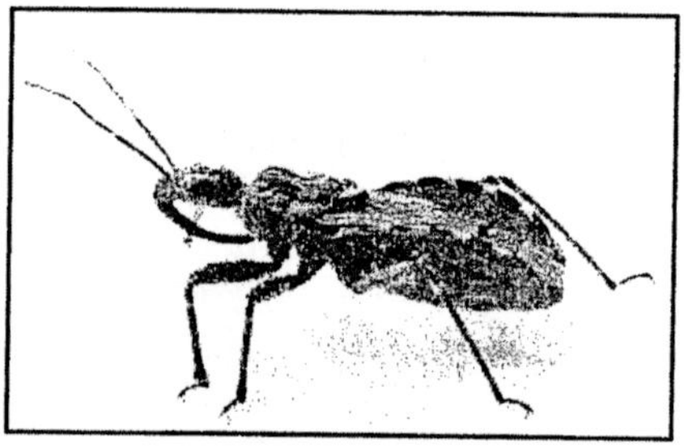

Many Hemiptera suck plant juices although some have evolved to suck blood and body fluids. Hemiptera groups include water scorpions, water boatman, backswimmers, water striders, plant bugs, bed bugs, assassin bugs, flat bugs, seed bugs, red bugs and stink bugs. Assassin bugs gain their sinister name from their habit of lying in ambush for their insect prey. With speed and precision, this bug injects its victim with a lethal toxin that dissolves tissue. Although many bugs use their sucking mouthparts for sipping plant juices, some have evolved as predators. When the food supply is limited, this insect will give up its ambush routine for a more active approach.

Assassin bug (family Reduviidae), any of about 4,000 species of insects in the true bug order, Heteroptera, that are characterized by a thin, necklike structure connecting the narrow head to the body. They range in size from 10 to 25 mm (0.4 to 1 inch). An assassin bug uses its short, three-segmented beak to suck the body fluids from its victims. A characteristic of this family is that the beak is curved and lies in a groove between the front legs. Although assassin bugs are generally black or dark brown, some species are brightly coloured. Most members of the family live outdoors and prey on other insects. However, some suck blood from vertebrates, including humans, and transmit diseases.

An important member of this family is the kissing bug (*Melanolestes picipes*). Its common name derives from the fact that it usually bites humans on the face around the mouth. This black-coloured insect is about 12 mm (0.5 inch) long and is usually found under stones and bark. It also feeds on other insects. The masked hunter (*Reduvius personatus*) is also known as the masked bedbug hunter, or kissing bug. During the immature stages the body, legs, and antennae are covered with sticky hairs that catch pieces of lint and dust, camouflaging the insect as a ball of dust. The brownish black adult, about 15 or 20 mm (0.6 or 0.75 inch) long, is commonly found in houses preying on insects, such as bedbugs and flies. It will also feed on humans and has a painful bite. Though originally a central European species, the masked hunter has spread throughout the southern parts of North America since its accidental introduction into the port of New York. One of the best-known assassin bugs is the cone-nose bug (*Triatoma*), also known as the kissing bug, big bedbug, or Mexican bedbug.

The adult is black with six red spots on each side of the abdomen and is about 25 mm long. The species *T. sanguisuga* is usually found in a bed, where it feeds on human blood. Its painful, toxic bite may cause faintness, swelling, and vomiting. The South American species *T. megista,* also known as the barber beetle, is the carrier of Chagas disease and a trypanosome protozoan disease. Chagas disease can also be transmitted by a related South American pest, *Rhodnius prolixus.* This bloodsucking assassin bug has been much used in insect physiology research. *T. rubrofasciata,* found living in the debris on hut floors in southern Asia, may transmit kala-azar. *Pristhesancus papuensis* is known as the bee killer.

This bug waits on flowers to capture and suck the body fluids from honeybees and other insects that frequent flowers. The wheel bug (*Arilus cristatus*) is recognized by the notched crest on the top of the thorax. The adult is gray and quite large (about 25 mm); the nymph is red with black marks. Wheel bugs occur in North America, are predaceous on other insects, and have a painful bite if handled. The venomous saliva is pumped into a victim through one channel in the wheel bug's beak.

The digested body fluids of the prey are then pumped into the wheel bug's stomach through another channel in the beak. The thread-legged bug *(Emesaya brevipennis)* is about 33 mm (1.3 inches) long and is usually found on trees or in old buildings. It has long, threadlike middle and hindlegs, while the shorter, thicker front legs are modified into viselike grasping organs. *E. brevipennis* occurs in North America.

The thread-legged bug *(Stenolemus bituberus),* which is native to Australia, preys on web-building spiders. It uses one of two different predatory strategies: stalking, in which it approaches its prey slowly and strikes when within range, or luring, in which it plucks the silk threads of the spider's web with its forelegs, which mimics the behaviour of an insect trapped in the web and thereby attracts the resident spider to within striking distance. The luring predatory behaviour displayed by *S. bituberus* represents a form of aggressive mimicry. An Asian member of the family, *Ptilocerus ochraceus,* has a most unusual method of capturing prey.

Tufts of red hair on its abdomen attract certain ants, which lick a glandular secretion from the hairs and become paralyzed. The assassin bug then pierces the ant with its beak and sucks out the body fluids. The large assassin bug *(Platymerus)* has an effective means of protection. Its powerful salivary pump enables it to "spit" saliva as far as 30 cm (12 inches). It has an accurate aim, and the saliva can cause blindness in humans. Some authorities also include the predatory ambush bug (Phymatidae) in the assassin bug family.

BED BUG

Kingdom: Animalia, **Phylum**: Arthropoda, **Class**: Insecta, **Order**: Hemiptera, **Suborder**: Heteroptera, **Infraorder**: Cimicomorpha, **Superfamily**: Cimicoidea, **Family**: **Cimicidae** Latreille, 1802.

Cimicidae or bedbugs, are small parasitic insects. The most common type is *Cimex lectularius*. The term usually refers to species that prefer to feed on human blood. All insects in this family live by feeding exclusively on the blood of warm-blooded animals. A number of health effects may occur due to bed bugs including skin rashes, psychological effects and allergic symptoms. Diagnosis involves both finding bed bugs and the occurrence of compatible symptoms. Treatment is otherwise symptomatic. Bed Bug, small, flat, oval bloodsucking insect, with a reddish-brown color, reduced wings, and an

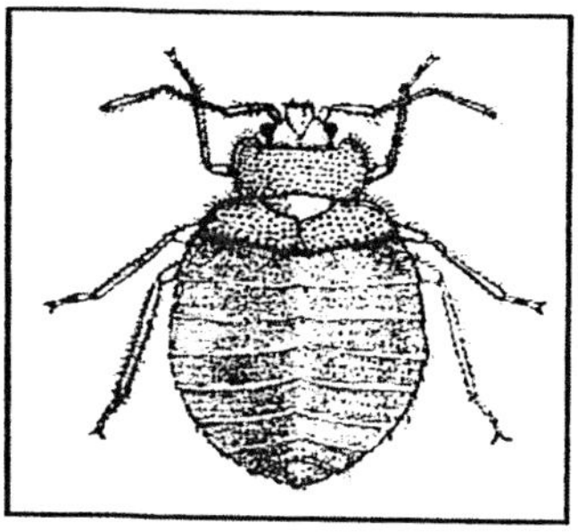

unpleasant odour. The common bed bug infests houses, particularly beds. It hides during the day and feeds at night, sucking the blood of humans and other mammals. Two to four generations are produced yearly, depend-ing on the temperature and the food supply. The nymphs resemble the adults, and they mature in about 11 weeks. Other species of bed bugs feed on bats and certain birds.

Habitat

The name "bedbug" is derived from the insect's preferred habitat of houses and especially beds or other areas where people sleep. Bedbugs, though not strictly nocturnal, are mainly active at night and are capable of feeding unnoticed on their hosts. They have however been known by a variety of names including wall louse, mahogany flat, crimson rambler, heavy dragoon, chinche and redcoat.

Life Stages

Bedbugs have six life stages (five immature and an adult stage). They will shed their skins through a molting process (ecdysis) throughout multiple stages of their lives. The discarded outer shells look like clear, empty exoskeletons of the bugs themselves. Bedbugs must molt six times before becoming fertile adults.

Threats

Although bed bugs can dine on any warm-blooded animal, they primarily dine on humans. Bed bugs do not transmit diseases, but their bites can become red, itchy welts.

Prevention

- Vacuum suitcases after returning from a vacation.
- Check your bedsheets for tell-tale blood spots.
- Consider bringing a large plastic trashbag to keep your suitcase in during hotel stays.
- Carry a small flashlight to assist you with quick visual inspections.
- Never bring second-hand furniture, especially mattresses and box springs, into a home without thoroughly examining for signs of a bed bug infestation. You might consider having a pest control professional inspect the furniture as it is difficult to detect an infestation if you are untrained.
- Regularly inspect areas where pets sleep for signs of bed bugs.
- Bed bugs are elusive creatures, so it is imperative to seek professional pest control to address an infestation.

CHINCH BUG

The chinch bug is classified as Blissus leucopterus. It belongs to the family Lygaeidae, order Heteroptera. Chinch bug small North American bug (*Blissus leucopterus),* of the seed bug family. It feeds on small grains, corn, and other grasses, sucking the plant juices and doing much damage to crops, particularly in the Midwest. The adults, about 1/8 in. (3.5 mm) long, have black bodies with black and white wings, red legs, and red spots at the bases of the antennae. Both long- and short-winged forms occur. There are two generations a year. The adults overwinter in sheltered places, emerging in spring to feed on early maturing grains, such as wheat and oats. They lay their eggs on the bases of the grasses or in the ground, and the nymphs, or larvae, emerge in about a week. Red when they emerge, the nymphs mature in five stages, turning gray or

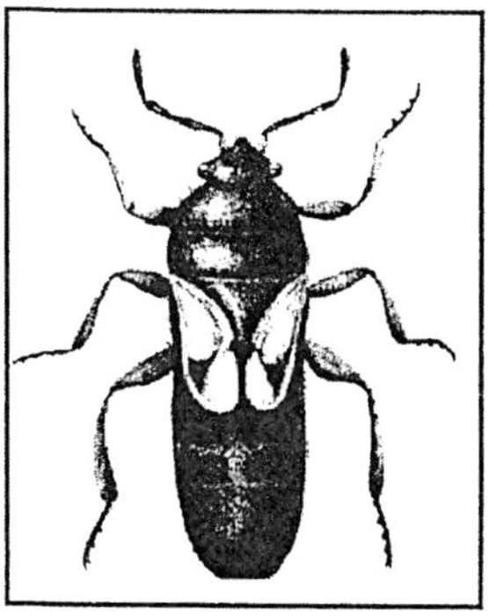

brown. They feed on the same grasses as their parents. When they reach the adult stage, in about six weeks, they migrate on foot to later-maturing grains, such as corn, which are still tender; there they lay the eggs that give rise to the second generation of the season. The bedbug , a member of a different bug family, is sometimes called chinch in the South. Chinch bugs are classified in the phylum Arthropoda , class Insecta, order Hemiptera, family Lygaeidae.

KISSING BUG

Kingdom: Animalia, **Phylum**: Arthropoda, **Class**: Insecta, **Order**: Hemiptera, **Family**: Reduviidae, **Subfamily**: **Triatominae** Jeannel, 1919.

Kissing bugs belong to the family Reduviidae, of the order Hemiptera. The big bed bug, or conenose, is classified as Triatoma sanguisuga. The South American barbeiro is classified as *Triatoma megistus*. A kissing bug is an insect in the subfamily *Triatominae,* part of the larger family *Reduviidae*. There are thousands of species in the Triatominae subfamily, most of which are found in the Americas, especially in tropical and subtropical regions. Kissing bugs are also known as assassin bugs, masked hunters, conenoses, and triatomines, and in some regions, they are a major pest.

The common name "kissing bug" is a reference to the feeding behavior of these insects. Kissing bugs feed on blood from humans and other animals, and they often feed from the area around the lips, where the skin is thinner, making it easier to penetrate. Kissing bugs are active at night, using their excellent sense of smell to track down victims while they sleep, and they typically alight next to their prey, stretching out a long proboscis to the area of interest. Triatomines have black to brown flattened bodies which are designed to swell as they

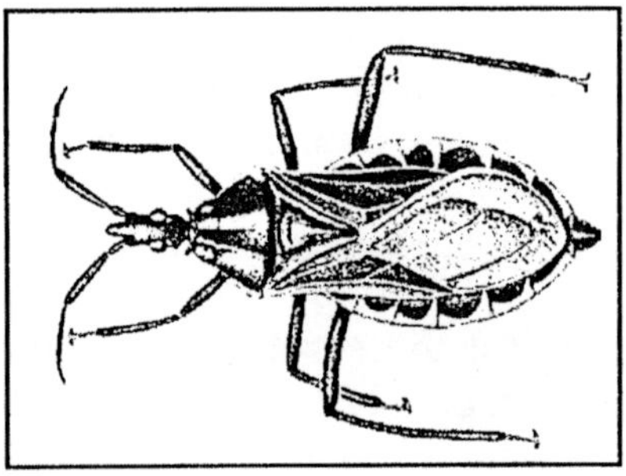

feed. In studies on these insects, researchers have discovered that the kissing bug can go up to four months between feedings. The insects commonly live in animal nests, taking advantage of a ready potential food supply, and they are indiscriminate feeders.

Triatoma infestans is one of the Kissing Bugs. Like the ticks these bugs can carry human diseases. *T.infestans* is the most important carrier of Chaga's disease in South America. What happens is that the bug gets infected with the parasite (*Trypanosoma cruzi*) when he sucks blood from an infected person, and then passes it on when he bites a new victim. The parasite is actually passed out with the bug's faeces, and then it creeps into the wound which is usually on the lips. Soon after infection there is a chronic reaction, typically with swelling around the lips, and then the disease lays dormant for several years. The later, chronic, stage of the disease is very serious. There is often severe heart damage, and while the disease can be treated just after infection, the chronic disease remains incurable. Even heart transplants don't seem to work because the parasite just takes advantage of the lowered immune system, reproduces like mad, and then starts doing damage all over again!

Charles Darwin (famous for his views on evolution) probably caught Chaga's Disease when he was bitten by *"a great black bug of the pampas"* while on his famous voyage round the world in the 'Beagle'. It is possible to get a vaccination before going to places where the disease is endemic (parts of South America), and you can also use mosquito nets. The worrying thing is that for every one adult flying down to 'get' you there will be five young ones crawling up! And that's not all one has young that camouflage themselves, so that you can end up carrying them into the house with the firewood.

Kissing Bugs are members of the Assassin Bug and Wheel Bug family of insects (Reduviidae). They have been found preserved in Amber - (Assassin Bug in Colombian Amber). Most Assassin Bugs kill insects so they can be very helpful to man, often killing agricultural and garden pests, and sometimes more directly (such as the 'Masked Hunter' who kills bedbugs!). Wheel Bugs are among the many bugs kept as pets by some people who feed them mealworms.

STINK BUG

Kingdom - Animalia. **Phylum** - Arthropoda. **Class** - Insecta. **Order** - Hemiptera. **Family** - Pentatomidae

The name Pentatomidae, the stink bug family, comes from the Greek "pente," meaning five and "tomos," meaning section. Some entomologists say this refers to the 5-segmented antennae, while others believe it refers to the stink bug's body, which appears to have five sides or parts. Either way, adult stink bugs are easy to recognize, with wide bodies shaped like shields. A long, triangular scutellum characterizes an insect in the family Pentatomidae. Take a close look at a stink bug, and you will see the piercing, sucking mouthparts.

Stink bug nymphs often resemble their adult counterparts, but may lack the distinctive shield shape. Nymphs tend to stay close to the egg mass when they first emerge, but soon venture out in search of food. Look for masses of eggs on the undersides of leaves.

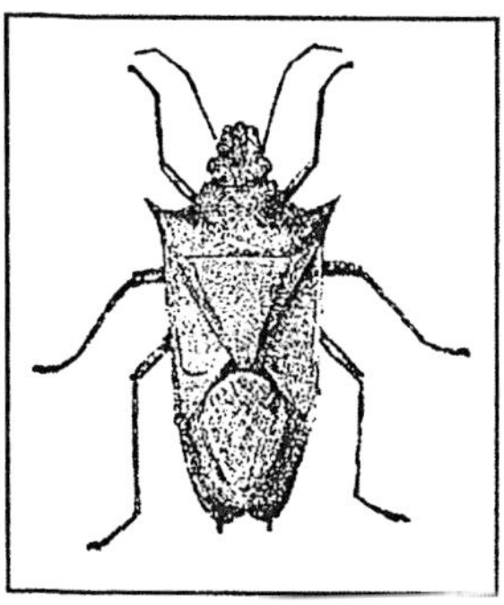

Stinkbug, common name applied to a bug characterized by a persistent disagreeable odour that protects the bug from enemies and that remains on plants over which the bug walks. The odour comes from two glands on the thorax. The family contains more than 5000 species, about 500 of which are native to the United States. Stinkbugs are broad, flat insects with five-jointed antennae. Their color is usually green, gray, and brown; some are coloured bright red and black or bright yellow and black. Stinkbugs vary in their food habits, some being exclusively carnivorous and others being exclusively plant feeders. One of the most familiar species of stinkbug is the very destructive cabbage bug.

Adults are approximately 17 mm long (25 mm = one inch) and are shades of brown on both the upper and lower body surfaces. They are the typical "shield" shape of other stink bugs, almost as wide as they are long. To distinguish them from other stink bugs, look for lighter bands on the antennae and darker bands on the membranous, overlapping part at the rear of the front pair of wings. They have patches of coppery or bluish-metallic coloured puntures (small rounded depressions) on the head and pronotum. The name "stink bug" refers to the scent glands located on the dorsal surface of the abdomen and the underside of the thorax.

The eggs are elliptical (1.6 × 1.3 mm), light yellow to yellow-red with minute spines forming fine lines. They are attached, side-by-side, to the underside of leaves in masses of 20 to 30 eggs.

WATER BOATMAN

Class: Insecta. **Order**: Hemiptera. **Suborder:** Heteroptera, **Family**: Corixidae (over 100 species).

Common Names

- Waterboatman, Boatman, Water Bug, Water Beetle, Paddle Bugs

Water Boatman, aquatic bugs that paddle along the water surface with oarlike hind legs. Water boatmen occur in fresh or brackish water throughout the world. In certain ponds or lakes they may be extremely abundant. About 525 species are known worldwide, 132 in North America.

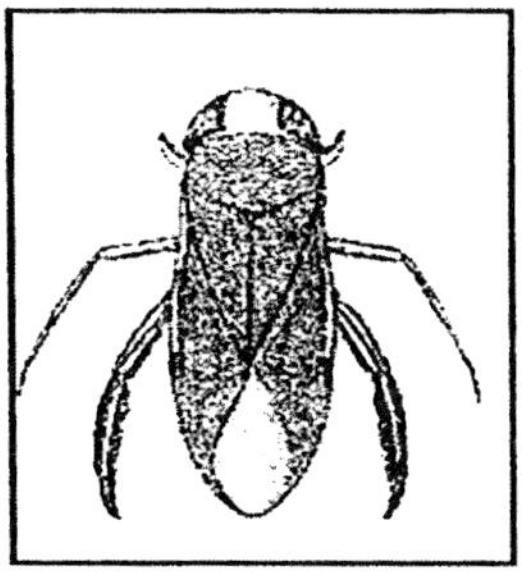

Waterboatmen spend most of their life in the water but do not have gills. When a tiny boatman nymph hatches from the egg it will swim to the water surface to get its first breath of air. It will also grab an air bubble with its modified legs and then swim back down to the lake bottom. Throughout its life, the boatman nymph will make many trips to the surface to replenish this air bubble for that is how it breathes. However, this bubble does act somewhat like a gill in that it absorbs oxygen from the water allowing the boatman nymph to stay submerged for long periods of time.

Over the course of the summer the nymph will feed on aquatic vegetation, algae and a few aquatic larva of other species and gradually mature into an adult. When the timing is right, the adult will swim to the surface, hesitate while the wings dry and then take flight. It will join a few, or many thousands of other Waterboatmen in a mating flight. The adults will mate in flight and then re-enter their aquatic home where the fertilized female lays her eggs on the submerged stems of aquatic vegetation. Water boatmen have a grayish, elongated, oval body, 3 to 12 mm (0.13 to 0.5 in) long. They have a conical beak and a broad head with large eyes. Adults have short, flattened front legs; long, slender middle legs; and oarlike hind legs fringed with fine hairs that aid in swimming.

Lifr Cycle

Eggs are attached to aquatic vegetation as they are laid. Nymphs develop through five growth stages, or instars, and have incomplete metamorphosis. Like all aquatic bugs, water boatmen lack gills; they breathe air when at the surface of the water. They frequently carry an air bubble on their body surface or under their wings, and draw oxygen from this bubble while they are underwater. Water boatmen can swim rapidly, but they spend long periods clinging to vegetation. Males stridulate, or chirp, to attract mates by rubbing their forelegs against their head.

Water boatmen occur most commonly in ponds and along the edges of lakes, although a few species inhabit the brackish waters of estuaries. Most water boatmen eat algae and minute aquatic organisms. Some are predaceous and feed on mosquito larvae and other small aquatic animals; in this way, they help to control aquatic pests. In turn, they are important prey for many larger aquatic animals. Their broad beak or mouth allows them to ingest solid food particles as well as liquids; other true bugs are able to ingest only liquids.

Unlike many other aquatic bugs, water boatman will not bite people. Water boatmen are sometimes confused with backswimmers, which are generally larger bugs that swim upside down and deliver a painful bite. Water-boatmen eggs are used as food in Mexico and some other parts of the world. Eggs are collected from aquatic plants, dried, and ground into flour.

WATER BUG

Kingdom: Animalia **Phylum**: Arthropoda **Class**: Insecta **Order**: Hemiptera **Infraorder**: Nepomorpha **Family**: Belostomatidae.

Water Bug, common name for aquatic insects of several families, including water boatmen, back swimmers, and water scorpions. The name water bug is sometimes restricted to the giant water bugs. This group of tropical to temperate species includes the largest of aquatic bugs, some of them more than 11 cm (more than 4.5 in) in length. Giant water bugs are also strong fliers. Their piercing, sucking mouthparts can deliver severe wounds through injection of proteolytic (protein-breaking) enzymes. Water boatmen, less than 12 mm (less than 0.5 in) in length, have long, oarlike hind legs. Water scorpions, which are approximately 2.5 to 5.0 mm (about 1 to 2 in) long, have long whiplike "tails" that are actually two respiratory tubes. Their painful bite is not as dangerous as the sting of true scorpions.

Life Cycle

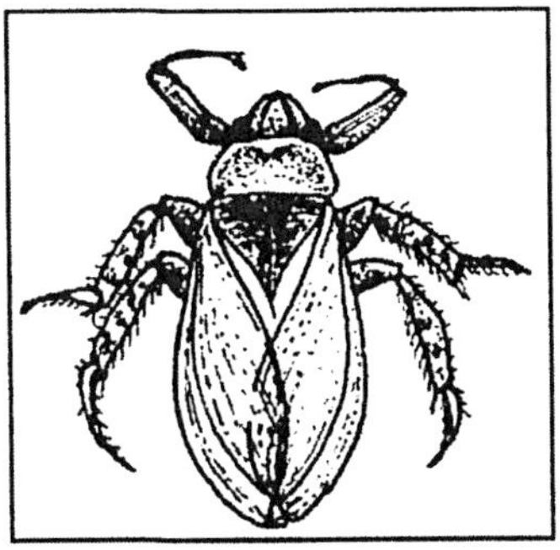

Males attract the females doing a series of periodic movements near water surface generating ripples in the water known as display pumping. An accompanying low frequency acoustical signal has been observed in at least one species of Belostomatinae. Before a female begins ovipositing the eggs, she mates with the male. Then a series of intercalated matings and ovipositions occur, females ovipositing 1-4 eggs in each ovipositing bout. An egg batch can have more than 100 eggs so a couple may copulate more than 30 times before female oviposits all the egg batch. This increases male confidence of paternity, a condition thought necessary for the origin of paternal care. There are two substrates of oviposition, females of species pertaining to the ancestral subfamily (Lethocerinae) oviposit on emergent vegetation. On the other hand, Belostomatinae females oviposit on the males' dorsum. The male will raise the eggs exposing them to air periodically to discourage the growth of fungus and maintain the viability of the eggs, or conduct a series of movements below the water known as brood pumping that increases the amount of oxygen diffusion. The eggs will hatch in approximately three weeks but their hatching time will decrease as temperature rises.

WATER STRIDER

Kingdom: Animalia **Phylum**: Arthropoda **Class**: Insecta **Order**: Hemiptera **Suborder**: *Heteroptera* **Infraorder**: Gerromorpha **Family**: Gerridae.

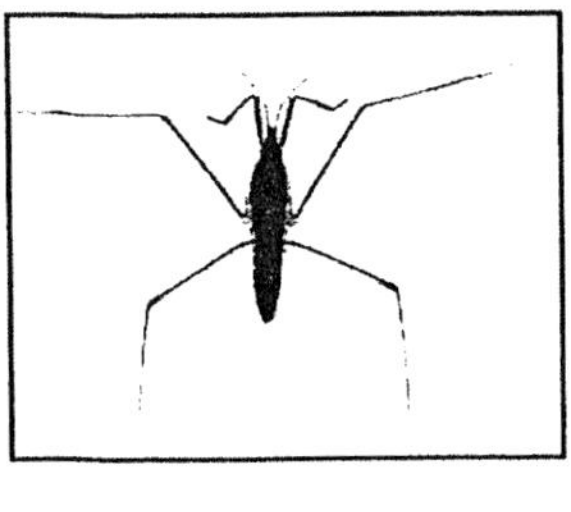

Water striders can vary in length from 1.6 mm (tiny species from the genus *Rheumatobates*) to 36 mm.

Similarly, their body shape ranges from slender and elongate to almost completely round. One common feature is their elongated legs (only the first pair is short and stubby) which the animals use for moving over the water surface. The body and legs are covered with tiny hairs. The head is usually elongated in front of the eyes to form a rostrum. Similar to other bug groups (such as Pyrrhocoridae), the development of wings can vary significantly within the same population. The population consists mostly of specimens with undeveloped or poorly developed wings. However, a small number of individuals have fully developed wings which they use for colonizing new habitats and forming new populations.

Butterflies and Moths

BUTTERFLY

Kingdom: Animalia, **Phylum**: Arthropoda, **Class**: Insecta **Order**: Lepidoptera, (**unranked**): **Rhopalocera.**

A butterfly is a mainly day-flying insect of the order Lepidoptera, the butterflies and moths. Like other holometabolous insects, the butterfly's life cycle consists of four parts, egg, larva, pupa and adult. Most species are diurnal. Butterflies have large, often brightly coloured wings, and conspicuous, fluttering flight. Butterflies comprise the *true butterflies* (superfamily Papilio-noidea), the *skippers* (super-family Hesperioidea) and the *moth-butterflies* (superfamily Hedyloidea). All the many other families within the Lepidoptera are referred to as moths.

Butterflies exhibit polymorphism, mimicry and aposematism. Some, like the Monarch, will migrate over long distances. Some butterflies have evolved symbiotic and parasitic relationships with social insects such as ants. Some species are pests because in their larval stages they can damage domestic crops or trees; however, some species are agents of pollination of some plants, and caterpillars of a few butterflies (e.g., Harvesters) eat harmful insects. Culturally, butterflies are a popular motif in the visual and literary arts.

Life Cycle

It is a popular belief that butterflies have very short life spans. However, butterflies in their adult stage can live from a week to nearly a year depending on the species. Many species have long larval life stages while others can remain dormant in their pupal or egg stages and thereby survive winters. Butterflies may have one or more broods per year. The number of generations per year varies from temperate to tropical regions with tropical regions showing a trend towards multivoltinism

Caterpillars

Butterfly larvae, or caterpillars, consume plant leaves and spend practically all of their time in search of food. Although most caterpillars are herbivorous, a few species such as *Spalgis epius* and *Liphyra brassolis* are entomophagous (insect eating). Some larvae, especially those of the Lycaenidae, form mutual associations with ants. They communicate with the ants using vibrations that are transmitted through the substrate as well as using chemical signals.

The ants provide some degree of protection to these larvae and they in turn gather honeydew secretions. Caterpillars mature through a series of stages called instars. Near the end of each instar, the larva undergoes a process called apolysis, in which the cuticle, a tough outer layer made of a mixture of chitin and specialized proteins, is released from the softer epidermis beneath, and the epidermis begins to form a new cuticle beneath. At the end of each instar, the larva moults the old cuticle, and the new cuticle expands, before rapidly hardening and developing pigment. Development of butterfly wing patterns begins by the last larval instar. Butterfly caterpillars have three pairs of true legs from the thoracic segments and up to 6 pairs of prolegs arising from the abdominal segments.

These prolegs have rings of tiny hooks called crochets that help them grip the substrate. Some caterpillars have the ability to inflate parts of their head to appear snake-like. Many have false eye-spots to enhance this effect. Some caterpillars have special structures called osmeteria which are everted to produce smelly chemicals. These are used in defense. Host plants often have toxic substances in them and caterpillars are able to sequester these substances and retain them into the adult stage. This helps making them unpalatable to birds and other predators. Such unpalatibility is advertised using bright red, orange, black or white warning colours. The toxic chemicals in plants are often evolved specifically to prevent them from being eaten by insects. Insects in turn develop countermeasures or make use of these toxins for their own survival. This "arms race" has led to the coevolution of insects and their host plants.

Wing Development

Wings or wing pads are not visible on the outside of the larva, but when larvae are dissected, tiny developing *wing disks* can be found on the second and third thoracic segments, in place of the spiracles that are apparent on abdominal segments. Wing disks develop in association with a trachea that runs along the base of the wing, and are surrounded by a thin *peripodial membrane,* which is linked to the outer epidermis of the larva by a tiny duct. Wing disks are very small until the last larval instar, when they increase dramatically in size, are invaded by branching tracheae from the wing base that precede the formation of the wing veins, and begin to develop patterns associated with several landmarks of the wing. Near pupation, the wings are forced outside the epidermis under pressure from the hemolymph, and although they are initially quite flexible and fragile, by the time the pupa breaks free of the larval cuticle they have adhered tightly to the outer cuticle of the pupa (in obtect pupae). Within hours, the wings form a cuticle so hard and well-joined to the body that pupae can be picked up and handled without damage to the wings.

MOTHS

Kingdom: Animalia, **Phylum**: Arthropoda, **Class**: Insecta, **Order**: Lepidoptera, (**unranked**): **Heterocera.**

A moth is an insect closely related to the butterfly, both being of the order Lepidoptera. Moths form the majority of this order; there are thought to be 150,000 to 250,000 different species of moth (about ten times the number of species of butterfly), with thousands of species yet to be described. Most species of moth are nocturnal, but there are crepuscular and diurnal species.

Differences between butterflies and moths

Moths are not easily differentiated from butterflies. Sometimes the name "Heterocera" is used for moths while the term "Rhopalocera" is used for butterflies to formalize the popular distinction; these, however, have no taxonomic validity. Many attempts have been made to subdivide the Lepidoptera into groups such as the Microlepidoptera and Macrolepidoptera, Frenatae and Jugatae, or Monotrysia and Ditrysia. Failure of these names to persist in modern classifications is because none of them represents a pair of "monophyletic groups". The reality is that butterflies are a small group that arose from within the "moths" (being considered as part of *Ditrysia* of the Neolepidoptera). There is thus no way to group all of the remaining taxa in a monophyletic group, as it will always exclude that one descendant lineage.

Characteristics

Different species of butterflies have different preferences of nectar, both in color and taste. A wide variety of food plants will give the greatest diversity of visitors. Try staggering wild and cultivated plants, as well as blooming times of the day and year. Groups of the same plants will be easier for butterflies to see than singly planted flowers Another way to attract adult butterflies to your yard is to offer places (food plants) for females to lay their eggs. Some females are pickier about which host to lay their eggs on than others. A few specific examples of butterflies and their host plants are listed at the end of this fact sheet. The larvae can also be very noteworthy. Some caterpillars have hairs or forked spines, which may be or may not sting (often the hairs are just for show).

It's better to be safe than sorry, so wear gloves when handling these larvae. Certain swallowtail caterpillars imitate snakes or bird droppings. Other caterpillars, like sulphers, are camouflaged, or blend into their surroundings very well. If caterpillars are eating excessive foliage from a prominent or desirable part of a plant, try moving them (with gloves on if they're hairy) to the backside or another less noticeable portion of the plant. Butterflies also like puddles. Males of several species congregate at small rain pools, forming puddle clubs. Permanent puddles are very easy to make by burying a bucket to the rim, filling it with gravel or sand, and then pouring in liquids such as stale beer, sweet drinks or water. Overripe fruit, allowed to sit for a few days is a very attractive substance (to them!) as well.

Controls

Some recent arrivals have become agents of biological control, selectively attacking weeds such as broom, wild parsnip or hemlock. Some species have been deliberately released to control troublesome plants. The gorse seed-pod moth (*Cydia succedana*) has been successfully established on gorse.

BUTTERFLY ANATOMY

Adult butterflies and moths have mouth parts shaped into a long, coiled tube. Forcing blood into the tube straightens it out,allowing butterflies to feed on liquids. Butterflies get all their food from this tube, which limits them to nectar and standing water. Larvae, on the other hand, have chewing mouth parts which they use to skeletonize or totally defoliate leaves. Butterflies have large, rounded compound eyes which allows them to see in all directions without turning their head. Like most insects, butterflies are very nearsighted, and are more attracted to large stands of a particular flower than those planted singly. They do not see "red" as well as we do, but they can see polarized light (which tells the direction the sun is pointing) as well as ultraviolet light, which is present on many flowers and guides them to nectar sources. Butterflies also have a very well-developed sense of smell from their antennae. All butterflies' antennae are club-shaped, as opposed to moths, which can be many shapes but often are feathery.

Butterflies begin their life as an egg, laid either singly or in clusters depending on the species. A very tiny caterpillar emerges and, after consuming its egg shell, begins feeding on its host plant. Caterpillars must crawl out of their skin or molt, usually around five times, before changing into a pupa. Finally, an adult butterfly emerges, spreads its wings and flies away. This type of development is complete metamorphosis.

Butterfly gardens are a great source for your own enjoyment, photo opportunities, or an outlet for artistic talent. These gardens can also be extended to interest youth in nature, by providing a small window of native inhabitants of the local environment. On a final note, it's important to conserve butterflies when possible since their habitat is constantly diminishing due to the increasing needs and consequent development of roads and housing.

Lepidopterans, i.e. butterflies and moths are insects, and like all insects their bodies are made up of three fundamental sections; the head, thorax and abdomen. On the head are two large compound eyes, two antennae and a long curled proboscis.It is a common misnomer that the antenna are feelers. Rather than a organ of touch they are an organ of smell. With butterflies the antenna are clubbed while in moths they are tapered. In some species of moths the antenna are enlarged into feathered organs. Increasing the surface area of the antenna produces a highly sensitive smell organ, particularly so with males. Being night animals seeking a mate visually is not an option, instead males seek by smell, picking up pheromones produced by females. Butterflies and moths feed on nectar, as seen in the image on the right. Nectar is obtained from flower heads by uncoiling and flattening the proboscis and then sucking the nectar up through it.The compound eyes of lepidopterans are large, they do not provide good resolution but are very good in sensing movement. Moths' visual range extends into the ultraviolet spectrum improving sight at night. However this does lead them to be drawn to light sources

Nutrition

Butterflies and moths undergo complete metamorphosis. Their life cycle consists of four stages: egg, larva (caterpillar), pupa (cocoon or chrysalis), and adult. After mating, the adult females of most species usually lay eggs on a plant that serves as the food source for the larvae when they hatch. In a few species, however, the larvae are predators. Some butterfly caterpillars, for example, eat aphids; some form complex associations with ants, live in their nests, and eat ant larvae. Some larvae eat stored cereals or even woolen clothes. Caterpillars have rather elastic cylindrical bodies, simple eyes, chewing mouthparts, three pairs of true legs on the thorax, and usually five pairs of prolegs (legs that occur only on larvae) on the abdomen. They eat continuously, periodically shedding their skin as they grow to hundreds of times their original size, and finally reach the stage where they spin cocoons and become pupae. During pupation, the structures of the larvae totally transform; internal systems are reorganized and adult external structures develop.

HABITAT

The butterfly habitat should be relatively sunny (5-6 hours per day) and out of the wind. Butterflies can get their required nutrients and minerals from a mud puddle in a sunny area. A successful butterfly habitat will contain the plants favored by both caterpillars and butterflies. This means that you should include plants like Queen Anne's Lace, violets, marigolds, and maybe even milkweed for the caterpillars to eat, since they are able to chew their food.

In particular, monarch caterpillars eat milkweed because it makes them an unpleasant-tasting meal to predators. For the butterflies, which are able to suck and are unable to chew, you should include nectar plants with different colours of flowers, such as zinnias, lilacs, and the butterfly bush. Butterfly habitats should last all summer, meaning that you should select plants that bloom at different times so that something is always blooming. You might be able to take inspiration in your planting selections from a butterfly garden or inside a butterfly house if you are lucky to live close enough to one of these carefully planned attractions. Often, landscapers and nursery employees will be able to advise you as to what possible additions to your butterfly habitat will grow well in your area. Butterflies and moths are found in a wide variety of habitats, from tundra to rain forest and from below sea level to nearly 6000 m (nearly 20,000 ft) in elevation. In tropical areas, where they reach their greatest diversity, many butterflies may fly throughout the year.

Continual good weather and ample resources allow for rapid larval development and long adult life, and as many as 15 generations may occur in one year. In temperate habitats, however, butterflies and moths enter an inactive stage, or diapause, during their development, to avoid severe weather conditions. Diapause may occur in the egg, larval, pupal, or adult stage. In snowy climates, winter hibernation is common; in areas with hot, dry seasons, summer diapause (estivation) is the rule. Larval development is generally slower in temperate areas, and an adult's life span is often only a few days or weeks. In alpine or arctic habitats, growing seasons may be so short that many species require two years for development.

ECOLOGICAL ROLES

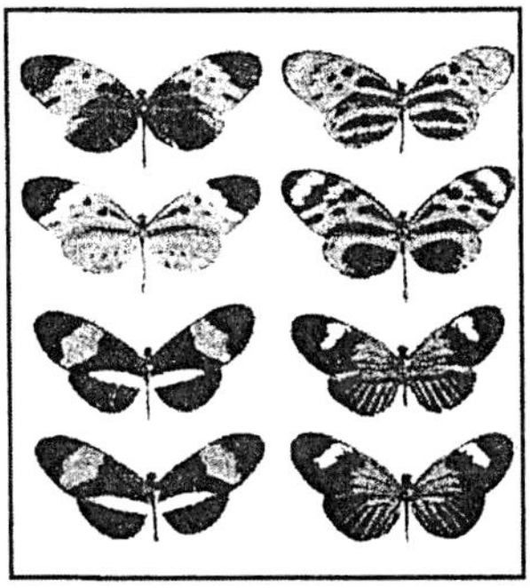

Butterflies and moths play several important ecological roles:

- Adults pollinate many plants whose flowers attract them with sugar-rich nectar.
- Larvae browse certain parts of plants, including seeds and flowers, influencing the plants' shape.
- Many predators and parasitoids eat them.

COLOURATION AND MIMICRY

The Lepidoptera, especially the butterflies, are known for the beautiful colours and patterns of their wings. Red, yellow, black, and white pigments are found in the scales; the blues and greens, and the metallic, iridescent hues found especially in tropical species, are caused chiefly by refraction. Some butterflies are protectively coloured to match the environment. Many conspicuously coloured species are distasteful to birds, which learn to avoid them, and others are protected by their resemblance to the distasteful species. Among the most beautiful butterflies are the swallowtails, found all over the world, the monarchs, and the peacock and tortoiseshell butterflies.

Mimicry, in biology, the advantageous resemblance of one species to another, often unrelated, species or to a feature of its own environment. (When the latter results from pigmentation it is classed as protective coloration.) Mimicry serves either to protect the mimic from its predators, as when the model is inedible or dangerous, or to deceive its prey (e.g., certain ant-eating spiders that themselves resemble ants). Mimicry occurs in both plants and animals but is most prevalent among insects, particularly butterflies and moths. The Batesian theory is based on the operation of natural selection: if, say, a harmless snake acquires a deceptive resemblance to a poisonous variety it is then more likely to escape its predators and thus to survive and propagate, producing offspring with the same appearance. Examples of mimicry are the resemblance of the viceroy butterfly to the monarch butterfly, which is repugnant to birds; harmless nettles that resemble stinging nettles; and the many fishes, crabs, and slugs of the Sargasso Sea that resemble the floating seaweed masses they inhabit. The Heliconius butterflies from the tropics of the Western Hemisphere are the classical model for Müllerian mimicry. Batesian and Mullerian mimicry in butterflies is common. Batesian mimics imitate other species to enjoy the protection of an attribute they do not share, aposematism in this case.

The Common Mormon of India has female morphs which imitate the unpalatable red-bodied swallowtails, the Common Rose and the Crimson Rose. Mullerian mimicry occurs when aposematic species evolve to resemble each other, presumably to reduce predator sampling rates, the Heliconius butterflies from the Americas being a good example. Wing markings called eyespots are present in some species; these may have an automimicry role for some species. In others, the function may be intraspecies communication, such as mate attraction. In several cases, however, the function of butterfly eyespots is not clear, and may be an evolutionary anomaly related to the relative elasticity of the genes that encode the spots.

IMPORTANCE OF BUTTERFLIES AND MOTHS

Butterflies,like bees, help cross pollinate plants. They get nectar from flowers which helps them with their pollination. Adult butterflies and moths are harmless, but their caterpillars (larvae) are often serious pests of plants, plant products, and certain animal products. Millions of dollars are spent each year

to combat injurious caterpillars. Some of the harmful species are the clothes moths, cabbage butterfly, corn borer, codling moth, gypsy moth, and cotton bollworm. There are more types of harmful moths than harmful butterflies.

A few kinds of butterflies and moths are important in the pollination of flowers. The hummingbird moth, for example, pollmates certain flowers while gathering nectar; the yucca moth places a ball of pollen in the yucca flower where it has just laid eggs. On the whole, however, butterflies and moths play a much smaller role in pollination than do bees. Butterflies are particularly useful for genetic experiments, are easily monitored in the field, and are thought to be representative of most herbivorous insects, humanity's most important competitors for food. Butterflies also serve as key indicator species of environmental disruption. Some moth species are among the most destructive pests of crops and stored products; however, very few butterfly species attack economically important plants. Many species of butterflies and moths are directly useful to humans. The silkworm moth is the primary example; its pupal cocoon is the source of commercial silk. Some species are eaten as protein-rich food, including silk moth pupae in Asia and many caterpillars in Africa.

BEE MOTH

A moth (*Galleria mellonella*) that lays its eggs in beehives, where the larvae feed on the wax and debris of the honeycombs. The bee moth belongs to the family Pyralidae.

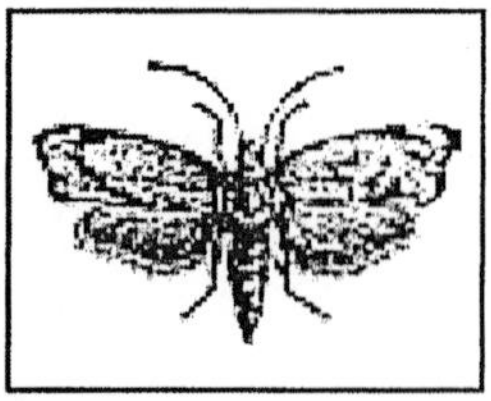

Also called wax moth. Bee Moth, common name for a moth having a slender, brownish or ashen-coloured body and long legs. The larva feeds on the wax of honey bee combs, often destroying the honey and injuring the bee brood. The larva of the smaller, but closely related, wax moth also feeds on honeycombs. bee moth, greater wax moth, or honeycomb moth, common name for an insect pest of honeycombs. Bee moths do damage during their larval stages, injuring combs and honey. The moth Galleria mellonella belongs to the subfamily Galleriinae of the family Pyralidae, in which the females characteristically lay their eggs in beehives. The adult moths have brownish front wings with wing-spans of about 1 in. (2.5 cm). Eggs are laid in masses in the crevices of the hive. The newly hatched larvae tunnel into the combs, leaving a complex of silken galleries behind; they also puncture the wax caps of honey cells causing honey leakage and making the punctured comb honey unmarketable. Normally, the moths attack only abandoned beehives, or active ones in which the bee colony has been weakened, e.g., as a result of disease or starvation. Another well-known but smaller member of the subfamily is the lesser wax moth, Achroia grisella, which has the same type of scavenging habits as the greater wax moth. Bee moths are classified in the phylum Arthropoda, class Insecta, order Lepidoptera, family Pyralidae, subfamily *Galleriidae.*

BROWNTAIL MOTH

Kingdom: Animalia, **Phylum**: Arthropoda, **Class**: Insecta, **Order**: Lepidoptera, **Family**: Lymantriidae, **Genus**: *Euproctis,* **Species**: *Euproctis* ***chrysorrhoea***

The Brown-tail (*Euproctis chrysorrhoea*) is a moth of the family Lymantriidae. It is distributed throughout Europe. The wings of this species are pure white, as is the body, apart from a tuft of brown hairs at the end of the abdomen. The brown colouration extends along most of the back of the abdomen in

the male. In the female, the back of the abdomen is white but the tuft of brown hairs is much bigger. Ths wingspan is 36-42 mm. The species flies at night in July and August and is attracted to light. The larva is brown with red and white markings and is very hairy. These hairs provide protection for this species throughout its life cycle: the larva incorporates some into the cocoon within which it pupates; the emerging adult female collects some on its tail and uses it to camouflage and protect the eggs as they are laid. The species overwinters communally as larvae within a tough silken tent. In areas where the species is abundant, these tents are a familiar sight and can be seen on a huge range of plants.

CABBAGE BUTTERFLY

Kingdom: Animalia, **Phylum**: Arthropoda, **Class**: Insecta, **Order**: Lepidoptera, **Family**: Pieridae, **Genus**: Pieris, **Species**: *Pieris rapae*

The body of the Cabbage Butterfly is black. All four wings are white. The two upper wings have a black band at the tips. Additionally, there is a black spot located in the center of each upper wing. There are also black spots on the lower wings, located in the center of each wing, too.The Cabbage Butterfly is 1.3-1.8 inches in width, and 1.5-2.0 inches in height. The caterpillar feeds on Cabbage.The butterfly is found all throughout North America, north into central Canada, and south into Mexico. The name is also loosely applied to several other related

species whose caterpillars feed on plants in the mustard family. The cabbage and other related plants secrete a chemical to which the butterfly is attracted. The butterfly deposits eggs on the leaves, on which the larvae feed and grow to maturity. The infested foliage is toxic upon consumption by animals. About a dozen species that follow these habits are found in North America, the most destructive of which is the imported cabbage butterfly, a European species introduced into Canada about the middle of the 19th century and now common throughout the United States. The fully grown larva, or caterpillar, is about 4 cm (about 1.5 in) long and green with a lemon-yellow dorsal band. It not only eats the leaves of the plants before heading time but also burrows through the heads. The butterfly has a wingspan of about 3 cm (about 1.2 in). The tops of its wings are white; the forewings are tipped with black. The forewings of the female have two central black dots, and those of the male, one. The undersides of the wings are pale yellow. Related species include the large white, a European variety; and the southern cabbageworm butterfly, or checkered white, a common American variety.

CATERPILLAR

It is widespread across Europe, North Africa and Asia and has also been accidentally introduced to North America, Australia and New Zealand where they have become pests on cultivated cabbages and other mustard family crops. Caterpillars are the larval form of a member of the order Lepidoptera (the insect order comprising butterflies and moths). They are mostly herbivorous in food habit, with some species being insectivorous. Caterpillars are voracious feeders and many of them are considered pests in agriculture.

Many moth species are better known in their caterpillar stages because of the damage they cause to fruits and other agricultural produce.

The etymological origins of the word are from the early 16th century, from Middle English *catirpel, catirpeller,* probably an alteration of Old North French *catepelose*: *cate,* cat (from Latin *cattus*) + *pelose,* hairy (from Latin *pilMsus*). Most caterpillars have tubular, segmented bodies. They have three pairs of true legs on the three thoracic segments, up to four pairs of prolegs on the middle segments of the abdomen, and often a single pair of prolegs on the last abdominal segment. There are ten abdominal segments.

The families of lepidoptera differ in the numbers and positioning of the prolegs. Some caterpillars are fuzzy (which means they have hair) and they are most likely to cause itching of the hands if touched. Caterpillars grow through a series of moults; each intermediate stage is called an instar. The last moult takes them into the inactive pupal or chrysalis stage. Like all insects, caterpillars breathe through a series of small openings along the sides of their thorax and abdomen called spiracles.

These branch into the body cavity into a network of tracheae. A few caterpillars of the family Pyralidae are aquatic and have gills that let them breathe underwater. Caterpillars have about 4,000 muscles (compare humans, with 629). The average caterpillar has 248 muscles in the head segment alone. Caterpillars move by sliding their internal organs forward using contraction of the muscles; the torso first elongates and then contracts like a slinky. They are one of the only known organisms that use this method of locomotion. Another way of describing this process is: gut-first, body-later.

ARMYWORM

Kingdom: Animalia, **Phylum**: Arthropoda, **Class**: Insecta, **Order**: Lepidoptera, **Family**: Noctuidae, **Genus**: *Spodoptera,* **Species**: *Spodoptera frugiperda.*

The Fall Armyworm (*Spodoptera frugiperda*) is part of the order of

Lepidoptera and is the caterpillar life stage of a moth. It is regarded as a pest and can wreak havoc with crops if left to multiply. Its name is derived from its feeding habits. Army worms will eat everything in an area and once the food supply is exhausted the entire "army" will move to the next available food source.

BAGWORM MOTHS

Kingdom: Animalia, **Phylum**: Arthropoda, **Class**: Insecta, **Order**: Lepidoptera, **Division**: Ditrysia, **Superfamily**: Tineoidea, **Family**: *Psychidae* Boisduval, 1828.

The common bagworm of the eastern United States is classified as *Thyridopteryx ephemeraeformis*. The species common in the southern United States is *Oiketicus abbottii*. The Psychidae (bagworm moths, also simply bagworms or bagmoths) are a family of the Lepidoptera (butterflies and moths). The bagworm family is fairly small, with about 600 species described. Bagworms species are found globally, with some, such as the Snailcase Bagworm (*Apterona helicoidella*), settling continents where they are not native in modern times. Another common name for the Psychidae is "case moths", but this is just as well used for the case-bearers (Colephoridae). The names refer to the habits of caterpillar of these two families, which build small protective cases in which they can hide. The bagworms belong to the superfamily Tineoidea, which is a basal lineage of the Ditrysia just as the Gelechioidea, in which the case-bearers are placed.

This means that the bagworms and case-bearers are only as closely related to each other as either is to butterflies (Rhopalocera). Most bagworms are inoffensive to humans and not at all conspicuous; some are occasional nuisance pests. However, a few species can become more serious pests, and have caused significant damage e.g. to wattle (*Acacia*) in South Africa and Orange (*Citrus × sinensis*) in Florida. If detected early, picking the cases from the trees while in their pupa stage is an effective way to check an infestation; otherwise, insecticides are used. One bagworm species, the Fangalabola (*Deborrea malgassa*) of Madagascar, is in some places encouraged to breed on wattle trees, because its pupae are collected as a protein-rich food.

BOLLWORM

Scientific name: *Helicoverpa armigera* (Synonym: Heliothis armigera), *Order/Family*: Lepidoptera: Noctuidae, *Type*: Pest (insect/mite), **Common names**: African bollworm, Fruitworm, Podborer, Corn earworm, Tomato grub, Tobacco budworm.

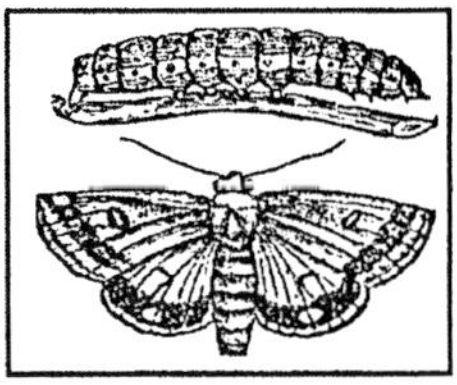

A larva of various moths (order Lepidoptera), including the pink bollworm of the family Gelechiidae and some *Heliothis* species. The larvae of the corn earworm (*Heliothis zea*), a member of the family Noctuidae, feed on the corn kernels near the tip of the ear. Because the larvae of this pest also burrow into tomatoes and cotton bolls, they are sometimes known as tomato fruitworms or cotton bollworms. It is classified as *Helicoverpa zea*. A bollworm is a common term for any larva of a moth that attacks the fruiting bodies of certain crops, especially cotton. The most common moths known as bollworms are:

- Red or Sudan Bollworm (*Diaparopsis*)
- Rough Bollworm (*Earias perhuegeli*)
- Spotted Bollworm (*Earias fabia*)
- Spiny Bollworm (*Earias insulana*)
- Spotted Bollworm (*Earias vittella*)
- American Cotton Bollworm or Tomato Grub (*Helicoverpa armigera*)
- Cotton Bollworm (*Helicoverpa gelotopoeon*)
- Cotton Bollworm (*Helicoverpa punctigera*)

- Corn Earworm (*Helicoverpa zea*)
- Tobacco Budworm (*Heliothis virescens*)
- Pink Bollworm (*Pectinophora gossypiella*)
- Pink Spotted Bollworm (*Pectinophora scutigera*).

CANKERWORM

Family: Geometridae, **classified as** *Paleacrita vernata,* the **fall cankerworm** as *Alsophila pometaria.*

The larvae of geometrid moths, that are destructive to fruit, buds and leaves. Cankerworm name for two destructive inchworms, or larvae of geometrid moths. The fall cankerworm (*Alsophila pometaria*) and the spring cankerworm (*Paleacrita vernata*) are named for the seasons at which the adults emerge from underground pupation. The spring cankerworm larva overwinters as a pupa, the fall cankerworm as an egg. The larvae, dark green to brown and about 1 in. (2.5 cm) long, feed on the leaves of orchard and shade trees. The spring cankerworm has two pairs of posterior appendages (prolegs); the fall cankerworm has three. The wingless female lays her eggs on the bark, and one control method is the placing of bands of sticky paper around the tree trunks to trap the females before laying. When alarmed, cankerworms drop and hang suspended in midair at the end of a long silken thread secreted from their mouths; they ascend this thread after the danger has passed.

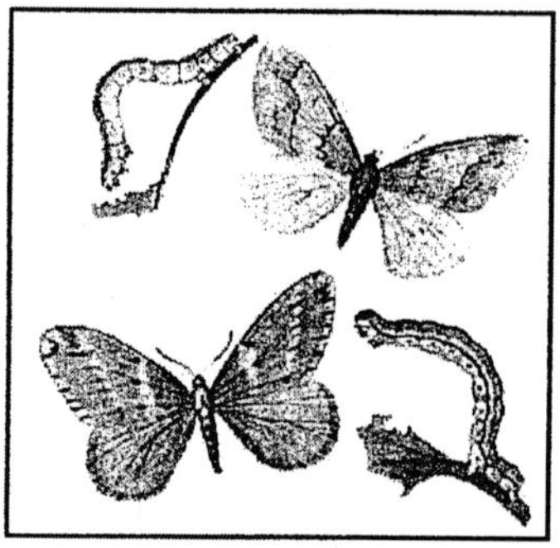

CUTWORM

The term **cutworm** is used for the larvae of the Turnip moth (*Agrotis segetum, Agrotis ipsilon, Agrotis exclamationis*). Most

cutworms are in the moth family Noctuidae including Agrotis, Euxoa, and several others, however, many noctuid larvae are not cutworms. It is classified as *Helicoverpa zea*. Cutworms are notorious agricultural and garden pests. They are voracious leaf, bud, and stem feeders and can destroy entire plants. They get their name from their habit of "cutting" off a seedling at ground level by chewing through the stem. Some species are subterranean and eat roots. Cutworms are usually green, brown, or yellow soft-bodied caterpillars, often with longitudinal stripes, up to 2.5 cm (1 inch) in length. There are many variations across the genera. The cutworms are mostly injurious to the conifers such as deodar pines etc. Cutworms also attack seedlings of casuarinas, teak, cassia and cashew.

The fertilized female moth lays about 2000 eggs in small lots on the stones, weeds, humus and other plants. After 26 days, hatching takes place after 26 days, and the larvae known as cutworms, feed on the green vegetation and dry leaves. In the day, cutworms hide in burrows of 30 to 75 mm deep in soil. At night, the larvae come out from the burrows to feed on the seedlings by cutting through the stem at ground level. The full grown larva is smooth, having short hairs which measures about 40 mm in length. Important species are: *Agrotis ipsilon, A. flammatro, A. segeturn, Laphygma exigua* and *Spodoptera litura*.

BEARDED FIREWORM

Kingdom: Animalia, **Phylum**: Annelida, **Class**: Polychaeta, **Family**: Amphinomidae, **Genus**: *Hermodice*, **Species**: *H. carunculata*.

Bearded fireworms are usually between 5–10 centimetres (1.9–3.9 in) in length, but can reach up to 35 centimetres (13.8 in). They are endowed with a group of poisonous white bristles on each side, which are flared out when the worm is disturbed. It is considered by marine biologists as a particularly beautiful

and colourful species, though swimmers wisely tend to avoid them due to their appearence.

The bearded fireworm is usually found on reefs, under stones in rocky areas of the sea, and on some mud bottoms. The black-headed fireworm is classified as *Rhopobota naevana*, and the yellow-headed fireworm as *Acleris minuta*. The marine fireworm is classified as Odontosyllis enopla in the order Errantia, class Polychaeta. Sometimes, they can be found to be hidden on moss-covered rocks. It is encountered throughout the tropical western Atlantic and at Ascension Island in mid-Atlantic. It can be found near ocean reefs and at depths of up to 150m. These worms are also common in the mediterranean sea, in the coastal waters surrounding Cyprus and the Maltese archipelago.

HORNWORM

Classified in the moth **family** Sphingidae, in the **order** Lepidoptera (butterflies and moths).

Hornworm, any of a family of moth caterpillars named for a prominent horn on the end of their abdomen. When disturbed, the hornworm will rear up and wave its horn in a threatening display to deter predators. Hornworms are harmless to humans, although a few species can damage agricultural crops. Hornworms live throughout the world and are most diverse in tropical regions.

There are about 1200 species worldwide, and 120 species in North America. Hornworms are among the largest of all

immature insects. Some grow to more than 10 cm (4 in) long and may weigh up to 10 gm (0.35 oz). Like many caterpillars, they are specialized herbivores that feed on leaves. Their digestive system often makes up more than half of the body, and some feed nearly continuously both day and night throughout their caterpillar lives, except when molting, or shedding their skin.

Under ideal conditions, a hornworm caterpillar may grow to weigh 10,000 times its original weight in less than two weeks. It then buries itself in leaf litter or under stones, and begins to transform into a pupa within the protective case of its larval skin. Tomato hornworms (*Manduca quinquemaculata*) are about the biggest caterpillars. They will startle you when you first spot them munching on your tomato plant. The tobacco hornworm is *Manduca sexta*.

These caterpillars feed ravenously on leaves and fruits of tomatoes and related plants. Adult hornworms are generally known as hawk or sphinx moths, although the term hornworm also applies to them. They have stout, fuzzy bodies, long, feathery antennae, and long, relatively narrow wings. Most hornworm moths are striped or patterned in shades of gray and brown, but some are quite strikingly coloured with black and red, green, or yellow. They feed on nectar through their long, hollow, tongue-like proboscises. Unlike butterflies, hornworms typically hover in the air while feeding at a flower, much as hummingbirds do, and for this reason are sometimes called hummingbird moths.

The unrolled proboscis extends into a long drinking straw capable of reaching nectar in flowers with narrow, deep corollas. These features enable the moths to feed on hanging, tubular flowers that are inaccessible to butterflies and to most bees. Moths of the hornworm family are powerful and fast fliers. Like many large insects, such as bumble bees, they maintain high body temperatures near 40° C (100° F) during flight. Under cold conditions, hornworm moths may shiver their flight muscles to generate heat before takeoff and flight.

This behaviour, known as thermoregulation, is unusual in insects. Because of their large size, their importance as agricultural pests, and the ease with which they can be raised in the laboratory, hornworms have become important experimental animals for studying insect physiology. Scientists have learned how hormones control molting and metamorphosis in insects from studies with hornworms, and they have used hornworms to study how arthropod nervous systems develop.

SILKWORM

Kingdom: Animalia, **Phylum**: Arthropoda, **Class**: Insecta, **Order**: Lepidoptera, **Family**: Bombycidae, **Genus**: *Bombyx*, **Species**: *Bombyx mori.*

The silkworm is the larva or caterpillar of the domesticated silkmoth (*Bombyx mori)* (Latin: "silkworm of the mulberry tree"). It is an important economic insect since it is the producer of silk. A silkworm's preferred food is white mulberry leaves, but it may also eat the leaves of any other mulberry tree (i.e., *Morus rubra* or *Morus nigra*) as well as the Osage Orange. It is entirely dependent on humans for its reproduction and does not occur naturally in the wild. Sericulture, the practice of breeding silkworms for the production of raw silk, has been underway for at least 5,000 years in China, from where it spread to Korea and Japan, and later to India and the West.

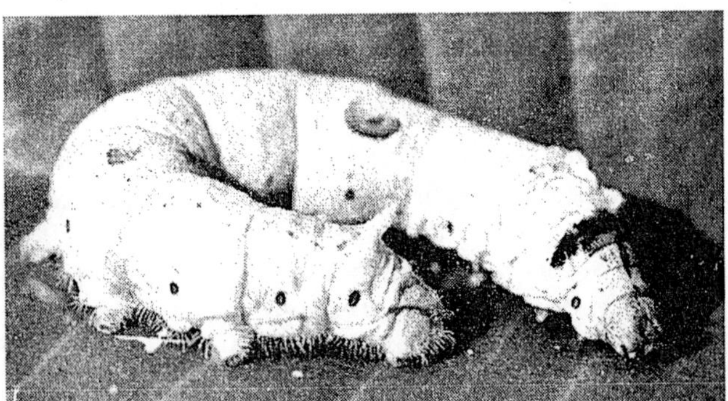

The silkworm was domesticated from the wild silkmoth *Bombyx mandarina* which has a range from northern India to northern China, Korea, Japan and far the eastern regions of Russia. The domesticated silkworm derives from Chinese rather than Japanese or Korean stock. It is unlikely that silkworms were domestically bred before the Neolithic age. It wasn't until then that the tools required to facilitate the manufacturing process of larger quantities of silk thread, had been developed. The domesticated *B. mori* and the wild *B. mandarina* can still breed and sometimes produce hybrids. Eggs take about fourteen days to hatch into larvae, which eat continuously. They have a preference for white mulberry, having an attraction to the mulberry odorant cis-jasmone. They are not monophagous since they can eat other species of *Morus* as well as some other Moraceae. Hatchlings and second-instar larvae are called kego and chawki in India.

They are covered with tiny black hairs. When the color of their heads turns darker, it indicates that they are about to molt. After molting, the instar phase of the silkworm emerges white, naked, and with little horns on the backs. After they have molted four times (i.e., in the fifth instar phase), their bodies become slightly yellow and the skin become tighter. The larvae will then enter the pupa phase of their life cycle and enclose themselves in a cocoon made up of raw silk produced by the salivary glands. The cocoon provides a vital layer of protection during the vulnerable, almost motionless pupal state.

Many other Lepidoptera produce cocoons, but only a few—the Bombycidae, in particular the Bombyx genus, and the Saturniidae, in particular the Antheraea genus—have been exploited for fabric production. If the animal is allowed to survive after spinning its cocoon and through the pupa phase of its life cycle, it will release proteolytic enzymes to make a hole in the cocoon so that it can emerge as a moth. These enzymes are destructive to the silk and can cause the silk fibers to break down from over a mile in length to segments of random length, and ruins the silk threads. To prevent this, silkworm cocoons are boiled. The heat kills the silkworms and the water makes the cocoons easier to unravel. Often, the silkworm itself is eaten. During the adult phase of the life cycle (the moth) cannot fly. Silkmoths have a wingspan of 3–5 cm (1.5–2 inches) and a white hairy body. Females are about two to three times the bulkier than males (for they are carrying many eggs), but are similarly coloured. Adult Bombycidaes have reduced mouth parts and do not feed, though a human caretaker can also feed them.

TENT CATERPILLARS

Kingdom: Animalia, **Phylum**: Arthropoda, **Class**: Insecta, **Order**: Lepidoptera, (**unranked**): Macrolepidoptera, **Superfamily**: Lasiocampoidea, **Family**: Lasiocampidae, **Subfamily**: Lasiocampinae, **Genus**: *Malacosoma.*

Tent caterpillars are moderately sized species in the genus *Malacosoma* and in the moth family Lasiocampidae. Species occur in North America, Mexico, and Eurasia. Twenty-six species have been described, six of which occur in North America. Some species are considered to have subspecies as well. They are often considered as pests due to their habit of defoliating trees.

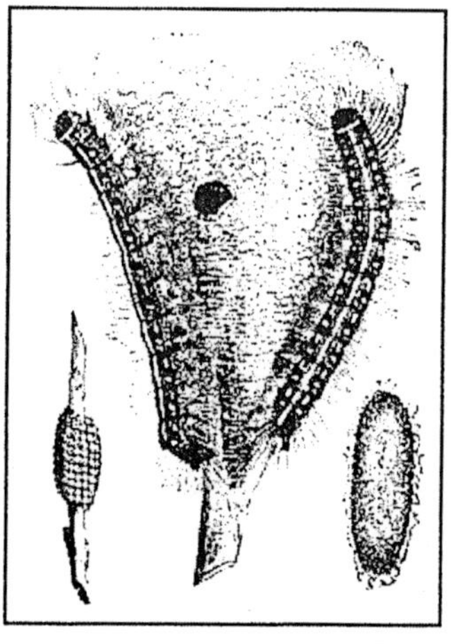

They are among the most social of all caterpillars and exhibit many noteworthy behaviours. Tent caterpillars are readily recognized because they are social, colorful, diurnal and build conspicuous silk tents in the branches of host trees. Some species, such as the eastern tent caterpillar, *Malacosoma americanum,* build a single large tent which is typically occupied through the whole of the larval stage while others build a series of small tents that are sequentially abandoned. The forest tent caterpillar, *Malacosoma disstrium*, is exceptional in that the larvae build no tent at all, aggregating instead on silken mats that they spin on the leaves or bark of trees.

Tents facilitate aggregation and serve as focal sites of thermal regulatory behaviour. They also serve as communication centers where caterpillars are alerted to the discovery of new food finds. Tent caterpillars should not be confused with Fall webworms or Gypsy Moths. Whereas tent caterpillars make their tents in the nodes and branches of a tree's limbs, webworms enclose leaves and small branches at the ends of the limbs in a light gray, silken webs. Tent caterpillars hatch from their eggs in the early spring at the time the leaves of their host trees are just unfolding. The caterpillars establish their tent soon after they eclose. The tent is constructed at a site that intercepts the early morning sun.

The position of the tent is critical because the caterpillars must bask in the sun to elevate their temperatures above the cool ambient temperatures that occur in the early spring. Studies have shown that when the body temperature of a caterpillar is less than about 15 °C, digestion cannot occur. The tent consists of discrete layers of silk separated by gaps and the temperature in these compartments varies markedly. Caterpillars can adjust their body temperatures by moving from one compartment to another. On cool mornings they typically rest in a tight aggregate just under a sunlit surface of the tent. It is not uncommon to find that the temperature of the aggregate is as much as 30 °C (54 °F) warmer than the surrounding air temperature on cold but sunny spring mornings. Later on in the spring, temperatures may become excessive at mid day and the caterpillars may retreat to the shaded outside surface of the tent to cool down.

WEBWORM

Kingdom: Animalia, **Phylum**: Arthropoda, **Class**: Insecta, **Order**: Lepidoptera, **Family**: Arctiidae, **Genus**: *Hyphantria*, **Species**: ***Hyphantria** cunea*.

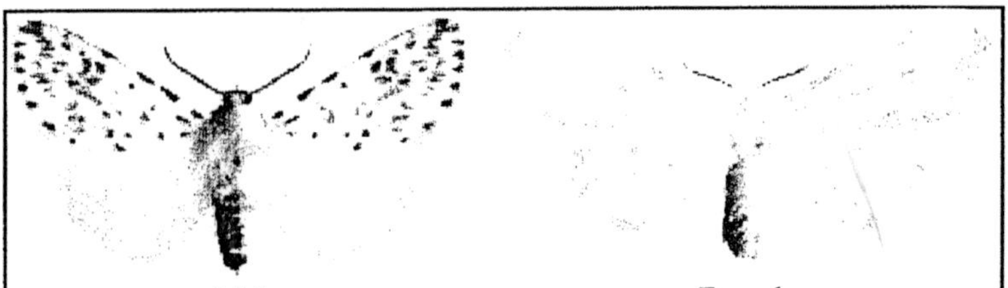

Male Female

Webworm (Hyphantria cunea), is a moth in the family Arctiidae known principally for its larval stage, which creates the characteristic webbed nests on the tree limbs of a wide variety of hardwoods in the late summer and fall. It is mainly an aesthetic pest and is not believed to harm otherwise healthy trees. It is well-known to commercial tree services and arboriculturists. One generation per year emerges in the northern part of North America, with larvae appearing in late summer through early fall. South of an approximate latitude of 40 degrees north there are two or more generations annually, with webs appearing progressively earlier further south. The adult moth lays her eggs on the underside of leaves in 'hair'-covered clusters of a couple of hundred. Eggs hatch in about a week.

The caterpillars are highly variable in coloration, ranging from a pale yellow, to dark grey, with yellow spots and long and short bristles. There are two cream stripes along the sides. There are two races, one more common in the north, the other in the south, differing in head capsule coloration. Maximum length, 35mm. Webs are progressively enlarged, and much messier looking than those of tent caterpillars (which occur only in spring and have shorter hairs and very little yellow on their bodies). Larvae feed inside the tents until the late instars. Very young larvae feed only on the upper surfaces of leaves. Later, they consume the whole leaf. The larval stage lasts about 4–6 weeks.

The pupal stage overwinters in the bark and leaf litter at the base of the trees. It is dark brown and about 10 mm long. They have thin brown cocoons made of silk with bits of detritus interwoven.

The adult is mostly white in the north, but in the south it may be marked with black or brown spots on the forewings. It is quite 'hairy', and the front legs have bright yellow or orange patches. The underwings will have less marking than the forewings, and the abdomen often has a sprinkling of brown hairs. Wingspan of 35-42 mm.

CECROPIA MOTH

Kingdom: Animalia, **Phylum**: Arthropoda, **Class**: Insecta, **Order**: Lepidoptera, **Family**: Saturniidae, **Genus**: Hyalophora, **Species**: *Hyalophora cecropia.*

The Cecropia Moth (*Hyalophora cecropia*) is North America's largest native moth. It is a member of the Saturniidae family, or giant silk moths. Females with a wingspan of 130 mm or more have been documented. It is found as far west as the Rocky Mountains and north into the maritime provinces of Canada. The larvae of these moths are most commonly found on Maple trees, but they have been known to feed on Wild Cherry and Birch trees among many others.

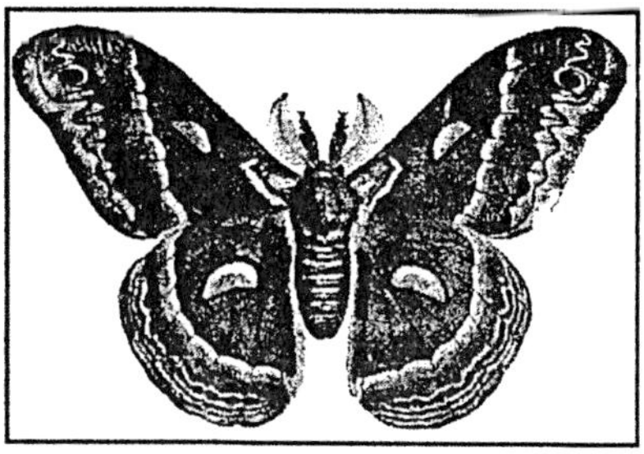

Life Cycle

The life cycle of the moth is much like any other saturniidae species. It lays oval-shaped, light brown eggs on the leaves of the host plant. When the eggs hatch, small black and hairy caterpillars emerge (first instar), and eat their eggshells. Sometimes the newly hatched caterpillars have yellow bristles, which darken in a few minutes. As the caterpillars age, they molt 4 times. The second instar is yellow-green with black spots and spiny orange, yellow, and blue knobs. The third and fourth instars have smaller spines, and are similar to each other. The fifth instar is a whitish color with bright yellow, blue, and red knobs.

The caterpillars feed heavily on their host plant and can grow up to five inches long. They then spin cocoons of brown silk, usually wrapped in leaves of the host plant. They then pupate inside the two-layered cocoon. There is only one brood per year, the cocoon overwintering and emerging in spring. When the adults eclose, they have to pump fluid into their wings to extend them. The females emit pheromones at night, which the male can detect with its large, plumose antennae. Males can fly for miles in order to reach a female. The moths mate, and the female spends the remainder of her life laying eggs, while the male may mate several more times. Because the adult moths do not have mouth parts they are unable to eat. As a result, the lifespan for an adult Cecropia moth is generally only 7 to 10 days in the wild, perhaps a little longer or shorter, depending on how much movement the moth makes. For instance, if temperatures are cooler, the moth may remain mostly inactive, which will increase its lifespan. In captivity, this lifespan can be extended to about a month by placing the moth in the fridge directly after it hatches and its wings are dry.

CHECKERSPOT

Kingdom: Animalia, **Phylum**: Arthropoda, **Class**: Insecta, **Order**: Lepidoptera, **Family**: Nymphalidae, **Genus**: *Euphydryas,* **Species**: *Euphydryas editha,* **Subspecies**: *Euphydryas editha bayensis.*

The Bay Checkerspot (*Euphydryas editha bayensis*) is a federally threatened insect subspecies which is native to the U.S. State of California. Since the 1980s the population of checkerspots has been in serious decline. Because this particular subspecies is so well studied, the decline was quickly recognized leading to its federally threatened status relatively quickly in 1987. Recently, two authors have advocated a reclassifi-

cation for this subspecies of the checkerspot common to western North America, *Euphydryas editha*. The classification, to *Euphydryas editha editha*, is for reasons of historical precedence; however, this has not been accepted by the scientific community.

Life Cycle

Adult butterflies emerge in early spring. The adults, which live an average of ten days emerge during a six week period from late February to early May. The male Bay checkerspot typically emerges four to eight days before the female. The males have one goal, reproduction. They find and mate with a female immediately upon emerging. The male mates many times while most females mate only once during the flight season. Besides mating the adults spend time foraging for nectar and for the females, laying eggs. Femalc checkerspots typically lay their eggs in clusters on a plant that serves as food for the larvae. Common food plants include plantain, beardtongue, Indian paintbrush, aster, rabbit brush, and other members of the figwort and sunflower families.

The eggs hatch in a few weeks and the young caterpillars that emerge feed together in a cluster; some species make a retreat of silk webbing similar to that of tent caterpillars. Checkerspot caterpillars are usually black or dark brown, sometimes with stripes of orange or other colours, and have ornamented spines along their backs. The caterpillars feed and molt several times over a period of several weeks. They spend the winter, curled up in a leaf or other protected site. In spring, they form a pupa inside a protective capsule, or chrysalis, and begin to change into adult butterflies. The speckled, camouflaged chrysalis hangs from a leaf or branch. The complete adult butterfly emerges in early summer and begins a new cycle.

If the butterflies within a particular patch become extinct, the empty patch will be recolonized after a period of time by butterflies moving from nearby patches. Although not all patches of habitat are always occupied by the butterflies, the butterfly population in the region as a whole persists as a result of this pattern of extinction and recolonization of local patches. A regional population may decline if occasional movement between the patches is prevented by disturbances such as land development. A population may also be in jeopardy if patches of habitats in a region are destroyed, even if that species does not currently occupy some of the habitat patches.

CLOTHES MOTH

Kingdom: Animalia, **Phylum**: Arthropoda, **Class**: Insecta, **Order**: Lepidoptera, **Family**: Tineidae, **Genus**: *Tineola*, **Species**: *Tineola bisselliella.*

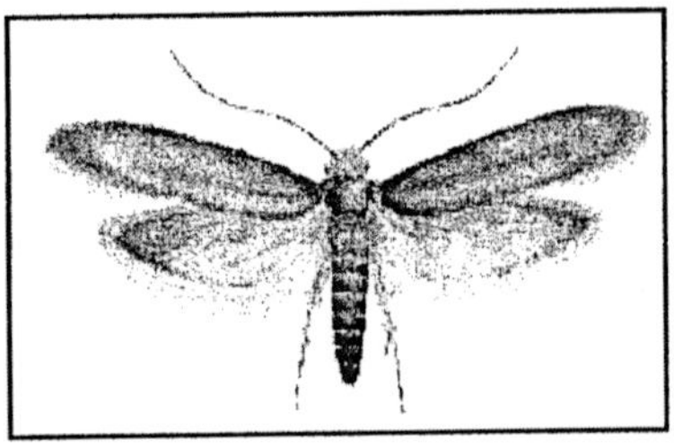

Tineola bisselliella, known as the Common Clothes Moth, Webbing Clothes Moth, or simply Clothing Moth, is a species of fungus moth (family Tineidae). Therein it belongs to the subfamily Tineinae. It is the type species of its genus *Tineola*. The specific name is commonly mis-spelled *bisellliella* – for example by G. A. W. Herrich-Schäffer, when he established *Tineola* in 1853. The caterpillars of this moth are considered a serious pest, as they can derive nourishment from clothing – in particular wool, but many other natural fibers – and also, like most moth of its relatives, from stored produce. Its larvae live within a case composed of bits of the food material bound together with silk. The larva of the carpet moth lives within a winding gallery made up of bits of carpet or other cloth held together with silk. Because clothes moth damage is easily avoided with the use of modern chemical repellents, these pests are now relatively uncommon.

CODLING MOTH

Kingdom: Animalia, **Phylum**: Arthropoda, **Class**: Insecta, **Order**: Lepidoptera, **Family**: Tortricidae, **Genus**: *Cydia*, **Species**: *Cydia pomonella.*

The codling moth (*Cydia pomonella*) is a member of the Lepidopteran family Tortricidae. They are known as an agricultural pest, their larva being the common apple worm or maggot. It is native to Europe and was introduced to North America, where it has become one of the regular pests of apple orchards. It is found almost worldwide. It also attacks pears, walnuts, and other tree fruits.

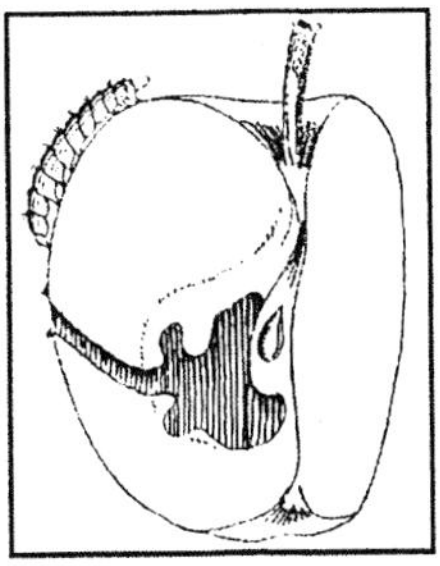

The larva is the infamous "worm in the apple" of cartoon and vernacular fame; it is not related to the earthworm. So it is also known as the apple worm. They also attack pears, quinces, and English walnuts. They are controlled by spraying the trees before the caterpillars have entered the fruit.

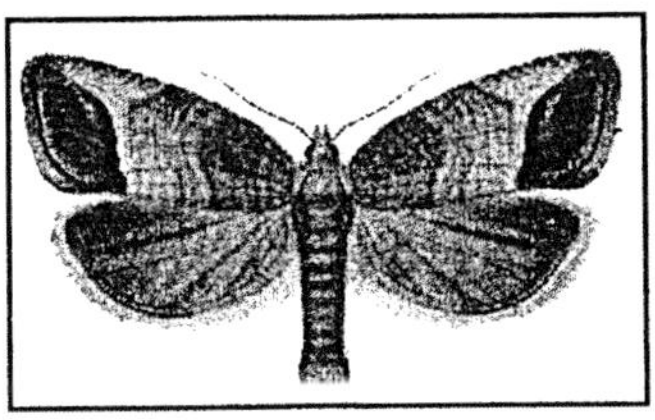

The codling moth is greyish with light grey and copper stripes on its wings, and has an average wingspan of 17 mm.

Life Cycle

The females lay eggs on fruit or leaves and the black-headed yellow larvae attack the fruit immediately upon hatching. Each larva burrows into the fruit, eats for around three weeks, then leaves the fruit to overwinter and pupate elsewhere. Most nourishment is obtained by feeding on the proteinaceous seeds. The codling moth is bivoltine in most regions of the USA - in the Pacific Northwest there is a partial third generation.

EMPEROR MOTH

Kingdom: Animalia, **Phylum**: Arthropoda, **Class**: Insecta, **Order**: Lepidoptera, **Family**: Saturniidae, **Genus**: *Pavonia*, **Species**: *Pavonia pavonia*.

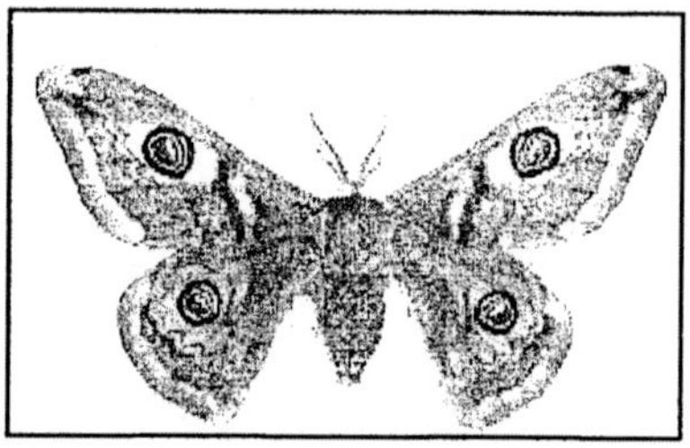

The Small Emperor Moth (*Pavonia pavonia*) is a moth of the family Saturniidae. It is sometimes placed in the genus *Saturnia*. It occurs throughout the Palearctic region and is the only member of its family to be found in the British Isles.

The male has a wingspan of about 60 mm with brown and white forewings marked with red and orange fascia and a bold black and orange eye-spot. The hindwings are orange with a similar eye-spot. The female is larger (wingspan about 80 mm) but less brightly coloured than the male, being generally grey and white but has all wings marked with eye-spots similar to the male.

The male flies rapidly during the day from mid-April to late June looking for the rather sluggish females, which usually only fly at night. The species inhabits a range of habitats but is most often associated with heathland and moorland.

The caterpillar is black and orange at first, later becoming green with black rings and yellow and red spots. The commonest food plant is heather but the species has also been recorded feeding on a huge range of other plants. The species overwinters as a pupa within a fibrous cocoon.

FRITILLARY

Kingdom: Plantae, **Division**: Magnoliophyta, (**unranked**): Monocots, **Order**: Liliales, **Family**: Liliaceae, **Subfamily**: Lilioideae, **Genus**: *Fritillaria* L.

Fritillaria is a genus of about 100 species of bulbous plants in the family Liliaceae, native to temperate regions of the Northern Hemisphere. The name is derived from the Latin term for a dice-box (*fritillus*), and probably refers to the checkered pattern, frequently of chocolate-brown and greenish yellow,

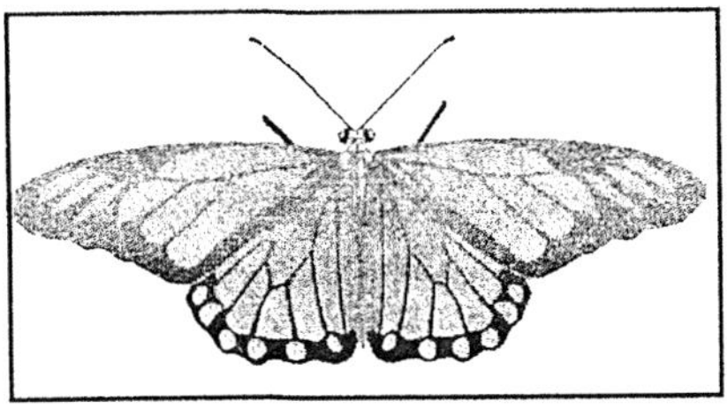

that is common to many species' flowers. Collectively, the genus is known in English as fritillaries; some North American species are called missionbells.

Wingspan of the fritillary ranges from 4 to about 10 cm. Fritillaries are typically vivid orange on the upper wing surfaces with black or brown markings. The undersides have rows of metallic silver or other spots. In most fritillary species the males and females are similar in wing color pattern, and females are typically larger than males. An exception is the Diana fritillary in which the males are orange and black above with silver spots below and females are iridescent blue-black with rows of white spots above, and charcoal black below. The colour pattern of female Dianas mimic those of a distantly related butterfly, the pipe vine swallowtail, which is distasteful to birds and other potential predators.

Life Cycle

Most fritillaries reproduce only once each year in temperate areas. In summer or early fall, the female will lay its eggs on or near violets, its host plant. Tiny caterpillars hatch from the eggs, and they spend the winter in a resting stage, called diapause, and resume growth and development in the spring. The mature caterpillar is typically black or dark brown with black or orange spines.

Another fritillary group, the lesser fritillaries, are similar in general wing pattern but are usually darker and smaller in size and live at northern latitudes in North America, Europe, and Asia. Some species are restricted to arctic and alpine regions where they are often seen basking with their hairy bodies and dark wings pressed against the ground or on rocks. Lesser fritillaries use a wider variety of host plants than other fritillaries, including violets, willows, bistort, and mountain avens.

GYPSY MOTH

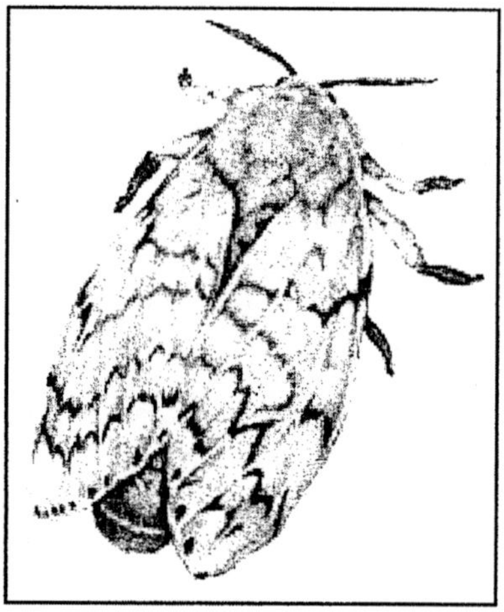

Kingdom: Animalia, **Phylum**: Arthropoda, **Class**: Insecta, **Order**: Lepidoptera, **Family**: Lymantriidae, **Genus**: *Lymantria*, **Species**: *Lymantria dispar.*

The gypsy moth, *Lymantria dispar*, is a moth in the family Lymantriidae of Eurasian origin. Originally ranging from Europe to Asia, it was introduced to North America in the late 1860s and has been expanding its range ever since. It is also known as the **Asian gypsy moth**.

Life Cycle

Gypsy moth egg masses are typically laid on branches and trunks of trees, but egg masses may be found in any sheltered location, particularly human dwellings, sometimes to the point of even nesting in clothes and footwear. Female gypsy moths in the U.S. are flightless, so they lay eggs on the surface where they emerged from their pupa or somewhere they can crawl - thus nearby. The egg is the overwintering stage. After an acclimation stage, eggs can withstand freezing temperatures. The longer they are chilled in winter, the less heating is required foı their hatch in spring. Egg clusters, usually an oval about 3/ 4 of an inch long, look like a piece of felt or velour and are buff coloured when first laid but may bleach out over the winter months when exposed to direct sunlight and weathering. As the female lays them, she covers them with hair-like setae from her abdomen. Many individuals find these hairs irritating, and they may offer the eggs some protection. Egg clusters contain from a couple of hundred to about 1,200 eggs, although some batches may be as small as 50 eggs.

Larvae develop into adults by going through a series of progressive moults through which they increase in size. Instars are the stages between each molt. Male larvae normally go through five instars (and females, six) before entering the pupal stage. Newly hatched larvae are black with long hair-like setae. Older larvae have five pairs of raised blue spots and six pairs of raised brick-red spots along their backs, and a sprinkling of setae. During the first three instars, larvae remain in the top branches or crowns of host trees.

The first stage or instar chews small holes in the leaves. The second and third instars feed from the outer edge of the leaf toward the center. When population numbers are sparse, the movement of the larvae up and down the tree coincides with light intensity. Larvae in the fourth instar feed in the top branches or crown at night. When the sun comes up, larvae crawl down the trunk of the tree to rest during daylight hours. Larvae hide under flaps of bark, in crevices, or under branches - any place that provides protection. When larvae hide underneath leaf litter, mice, shrews, and *Calosoma* beetles can prey on them. At dusk, when the sun sets, larvae climb back up to the top branches of the host tree to feed. When population numbers are dense, however, larvae feed continuously day and night until the foliage of the host tree is stripped. Then they crawl in search of new sources of food. Full-grown caterpillars are 40–50 mm long.

The larvae reach maturity between mid-June and early July. They enter the pupal stage. This is the stage during which larvae change into adults or moths. Pupation lasts from 7 to 14 days. When the population is spread out and running low, pupation can take place under flaps of bark, in crevices, under branches, on the ground, and in other places where larvae rested. During periods when population numbers are dense, pupation is not restricted to locations where larvae rested. Pupation will take place in sheltered and non-sheltered locations, even exposed on the trunks of trees or on foliage of nonhost trees. Usually the caterpillars create flimsy cocoons made of silk strands holding the leaf together, while others do not cover their pupae in cocoons, but rather hang from a twig or tree bark, like butterfly pupae do.

The forewing of the male moth is 20–24 mm long, and that of the female 31–35 mm. The brown male gypsy moth emerges first, flying in rapid zigzag patterns searching for females. The male gypsy moths are active throughout night and even daytime as well, unlike most moths, which are only nocturnal. When heavy, black-and-white egg-laden females emerge, they emit a pheromone that attracts the males. After mating, the female lays her eggs in July and August close to the spot where she pupated. Then, both adult gypsy moths die. The European and most Russian forms of the gypsy moth have flightless females. Although they have large wings, the musculature is not developed. However, the Japanese gypsy moth females do fly and are attracted to lights. Gypsy moths fly all day and night, with the possible exception of the late morning. They are most active soon after dusk and in the latter hours of the night

LUNA MOTH

Kingdom: Animalia, **Phylum**: Arthropoda, **Class**: Insecta, **Order**: Lepidoptera, **Family**: Saturniidae, **Genus**: *Actias*, **Species**: *Actias luna*.

The Luna moth (*Actias luna*) is probably the best-known of the American silkmoths. Although it doesn't produce a large amount of silk, and isn't really practicable for textile use, I am rearing it for sake of its beautiful green wings.

Based on the climate in which they live, the Luna Moths produce differing numbers of generations. In Canada and northern regions, they can live up to 7 days and will produce only one generation per year. These reach adulthood from early June to early July. In the northeastern United States around New Jersey or New York, the moths produce two generations each year. The first of these appear in April and May, and the second group can be seen approximately nine to eleven weeks later. In the southern United States, there can be as many as

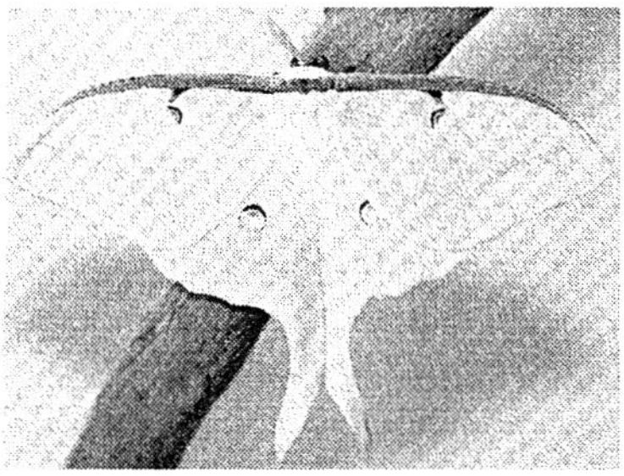

Male

Female

three generations. These are spaced every eight to ten weeks beginning in March.

Life Cycle

Female Luna Moths lay 100–300 eggs, 4–7 eggs at a time, on the underside of leaves, and they incubate for eight to thirteen days. The moths will lay more eggs in a favorable climate.

Each instar generally takes about five days to a week to complete. After hatching, the caterpillars tend to wander around before finally settling on eating the particular plant they are on. These caterpillars tend to be gregarious for the first two to three instars, but separate and live independently after that. As with all Saturniids, these caterpillars go through five instars before cocooning. At the end of each instar, a small amount of silk is placed on the major vein of a leaf and the larva then undergoes apolysis. The caterpillar then undergoes ecdysis, or molts from that position leaving the old exoskeleton behind. Sometimes the shed exoskeleton is eaten. Each instar is green, though the first two instars do have some variation in which some caterpillars will have black underlying splotches on their dorsal side. Variation after the second instar is still noticeable, but slight. The dots that run along the dorsal side of the caterpillars vary from a light yellow to a dark magenta. The final instar grows to approximately nine centimeters in length.

The Luna Moth pupates after spinning a cocoon. The cocoon is thin and single layered. Shortly before pupation, the final, fifth instar caterpillar will engage in a "gut dump" where any excess water, food, feces, and fluids are expelled. The caterpillar will also have an underlying golden reddish brown color and become less active. As a pupa, this species is particularly active. When disturbed, the moth will wiggle within its pupal case, producing a noise. Pupation takes approximately two weeks unless the individual is diapausing. The mechanisms for diapause are generally a mixture of genetic triggers, duration of sunlight or direct light during the day, and temperature.

Adults eclose, or emerge from their cocoons in the morning. Their wings are very small when they first emerge and they must enlarge them by pumping bodily fluids through them. During this time, their wings will be soft and they must climb somewhere safe to wait for their wings to harden before they can fly away. This process takes about 2 hours to complete. The Luna Moth has a wingspan of 8–11.5 cm with long, tapering hindwings, which have eyespots on them in order to confuse potential predators. Although rarely seen due to their very brief (1 week) adult lives, Luna Moths are considered common. As with all Saturniidae, the adults do not eat or have mouths. They emerge as adults solely to mate, and as such, only live approximately one week. They are more commonly seen at night.

MONARCH BUTTERFLY

Kingdom: Animalia, **Phylum**: Arthropoda, **Class**: Insecta, **Order**: Lepidoptera, **Superfamily**: Papilionoidea, **Family**: Nymphalidae, **Subfamily**: Danainae, **Tribe**: Danaini, **Genus**: *Danaus* Kluk, 1802, **Species**: *Danaus plexippus.*

The Monarch butterfly *(Danaus plexippus)* is a milkweed butterfly (subfamily Danainae), in the family Nymphalidae. It is perhaps the best known of all North American butterflies. Since the 19th century, it has been found in New Zealand, and in Australia since 1871 where it is called the Wanderer. In Europe it is resident in the Canary Islands, the Azores, and Madeira, and is found as an occasional migrant in Western

Male

Female

Europe. Its wings feature an easily recognizable orange and black pattern, with a wingspan of 8.9–10.2 centimetres.

Female Monarchs have darker veins on their wings, and the males have a spot called the "androconium" in the center of each hind wing from which pheromones are released. Males are also slightly larger. The Monarch is famous for its southward migration and northward return in summer in the Americas which spans the life of three to four generations of the butterfly. The Monarch's wingspan ranges from 8.9-10.2 cm.

The upper side of the wings is tawny-orange, the veins and margins are black, and in the margins are two series of small white spots. The fore wings also have a few orange spots near the tip. The underside is similar but the tip of the fore wing and hind wing are yellow-brown instead of tawny-orange and the white spots are larger.

The male has a black patch of androconial scales responsible for dispersing pheromones on the hind wings, and the black veins on its wing are narrower than the female's. The male is also slightly larger. A colour variation has been observed in Australia, New Zealand, Indonesia and the United States as early as the late 19th century. Named *nivosus* by Lepidopterists, it is grayish white in all areas of the wings that are normally orange.

Generally it is only about one per cent or less of all monarchs, but has maintained populations as high as 10 per cent on Oahu in Hawaii, possibly due to selective predation. Like all insects the Monarch has six legs, however it uses only four of its legs as it carries its two front legs against its body.

The eggs are creamy white and later turn pale yellow. They are elongate and subconical, with approximately 23 longitudinal ridges and many fine traverse lines. A single egg weighs about 0.46 milligrams (0.0071 gr), and measures about 1.2 mm high and 0.9 mm.

The caterpillar is banded with yellow, black, and white stripes. The head is also striped with yellow and black. There are two pairs of black filaments, one pair on each end of the body. The caterpillar will reach a length of 5 cm.

The chrysalis is blue-green with a band of black and gold on the end of the abdomen. There are other gold spots on the thorax, the wing bases, and the eyes.

PAINTED LADY

Kingdom: Animalia, **Phylum**: Arthropoda, **Class**: Insecta, **Order**: Lepidoptera, **Family**: Nymphalidae, **Subfamily**: Nymphalinae, **Tribe**: Nymphalini, **Genus**: *Vanessa*, **Subgenus**: *Cynthia.*

The Painted Lady group of colourful butterflies comprises the subgenus *Cynthia* of the genus *Vanessa* in the Family Nymphalidae. They are well known throughout most of the world. The group includes:

- the Painted Lady (*Vanessa cardui*), which is almost global in its distribution
- Australian Painted Lady (*Vanessa kershawi*)
- American (Painted) Lady (*Vanessa virginiensis*)
- West Coast Lady (*Vanessa annabella*).

In general, the Painted Lady is a large butterfly (wing span 5–9 cm (2.0–3.5 in)) identified by the black and white corners of its mainly deep orange, black-spotted wings. It has 5 white spots in the black forewing tips and while the orange areas may be pale here and there, there are no clean white dots in them. The hindwings carry 4 small submarginal eyespots on dorsal

and ventral sides. Those on the dorsal side are black, but in the summer morph sometimes small blue pupils are present in some.

The American Painted Lady (*V. virginiensis*) is most easily distinguishable by its two large hindwing eyespots on the ventral side. *V. virginiensis* also features a white dot within the subapical field of the forewings set in pink on the ventral side, and often as a smaller clean white dot in the orange of the dorsal side too. A less reliable indicator is the row of eyespots on the dorsal submarginal hindwing; *virginiensis* often has two larger outer spots with blue pupils. The black forewing tips have 4–5 white spots, usually the largest is whitish orange.

The West Coast Lady (*V. annabella*) does not have obvious ventral eyespots. On the dorsal side, *anabella* lacks a white dot in the subapical orange found in *virginiensis*, and is a purer orange colour. *V. annabella* has a fully orange subapical band and leading edge on the forewing. The submarginal row of hindwing spots in *annabella* features three or four blue pupils. The two larger pupils in *annabella* are the inner spots, rather than the outer spots as in corresponding *virginiensis*.

The Australian Painted Lady (*V. kershawi*) is quite similar to *V. cardui*. Its four ventral eyespots are less clearly defined, and it always sports at least three (often four) blue pupil spots on its dorsal hindwing. Caterpillars are found mainly on *Ammobium alatum*.

Unlike many butterflies, it does not become dormant during the winter and cannot survive heavy frosts or colder conditions; thus it must spend its winters in warmer climates. The painted lady feeds on nectar from the flowers of thistles, red clover, asters, and many other plants. The painted lady is famous for its extensive mass migrations. During the spring it migrates northward, sometimes in huge numbers, flying 3 to 4 m (10 to 14 ft) above the ground.

Life Cycle

The painted lady has four life stages: egg, caterpillar (wormlike larva), cocoon, and winged adult. The female lays green eggs one at a time on the leaves of plants that will later provide food for the caterpillar, such as thistles, mallows, or peas. The increasing abundance of thistles in ecologically disturbed areas has likely contributed to the increasing abundance of painted ladies. In some regions, the caterpillars are minor pests on lowers, soybeans, sunfand other crops. The caterpillars are covered with short spines and quite variable in colour.

PEACOCK BUTTERFLY

Kingdom: Animalia, **Phylum**: Arthropoda, **Class**: Insecta, **Order**: Lepidoptera, **Family**: Nymphalidae, **Subfamily**: Nymphalinae, **Tribe**: Nymphalini, **Genus**: *Inachis* Hübner, 1819, **Species**: *Inachis io.*

The Peacock is found throughout the U.K., though it is less common in parts of Scotland. However, due to climate change, it is moving further north into the Highlands and North-east Scotland. It inhabits gardens, parks, woodlands and any area where nettles, the favourite food of its caterpillars, grow. Adult butterflies emerge from hibernation in March and April, though a few may be spotted on warm days even in January. These butterflies pupated the previous summer and therefore are among the longest-lived U.K. butterflies. They feed generally in the mornings and search for territories in the afternoon, preferring sunny corners of woodland or park.

The scientific name, or two-part Latin binomial, for the Peacock Butterfly is *Inachis io*. The name Peacock Butterfly comes from the large eye spots on the wings, which look like those on the tail of a peacock. Peacock Butterflies are herbivores.

Adult Peacock butterflies feed on nectar, but the caterpillars eat nettles. Male Peacock butterflies are strongly territorial and will chase competing males from their patch. Any female

butterfly is hotly pursued. After mating, the females lay their eggs in May on nettles in clusters of up to 500. Eggs are laid in large, irregular clusters on the undersides of young nettle leaves. In contrast to the laying sites of the Small Tortoiseshell, large vigorous plants in the middle of nettle beds are often selected, although these must be in full sunlight at midday, when egg laying usually occurs. The adults then die and no further Peacock butterflies are seen until the new generation emerges in July. Peacock caterpillars are black with short spines and hairs. After hatching, they congregate to build a web over the growing tip of the nettle plant, moving on to other nettles as they strip the plant bare. They disperse to pupate and emerge as butterflies in July.

Distribution

The Peacock is widespread and common in the south of Britain and Ireland, but less frequent to the north and at higher altitudes. In Scotland, it is resident in the west and on islands of the west coast, as far north as Argyll. It is probably also resident in the south-east of Scotland. Scattered modern and historical records show that the butterfly has reached many northern parts of Scotland, including Orkney and Shetland, but these individuals are probably vagrants. In January 2004, a Peacock was found in hibernation in a stack of cut peat on the island of Lewis in the Outer Hebrides!

Little change in distribution occurred in the southern part of its range during the twentieth century. However, in the north, the Peacock expanded its range from an apparent low point in the latter part of the nineteenth century. This expansion began in the 1930s and has continued to the present day in some places, notably Northern Ireland. Here the butterfly was recorded in only one-quarter of 10 km squares in 1970-82, but in over three-quarters of squares in the period 1995-9. The butterfly is more widely recorded now across the Pennines and has become much more common in the central belt and highlands of Scotland in recent years.

As with many butterflies that are currently expanding their range, data from the Butterfly Monitoring Scheme show a significant overall increase in the abundance of the Peacock since 1976. This trend is strong despite dramatic population crashes following wet and dull summers in 1985 and 1986, which caused extinction at several northern BMS sites and scarcity in many areas, for example around Dublin. Further falls in the collated index, such as in 1993 and 1994, did not prevent a record value being achieved in 1996. As with the Small Tortoiseshell, the largest increases in abundance have been in eastern England.

RED ADMIRAL

Kingdom: Animalia, **Phylum**: Arthropoda, **Class**: Insecta, **Order**: Lepidoptera, **Family**: Nymphalidae, **Genus**: *Vanessa,* **Species**: *Vanessa atalanta.*

Red Admiral (*Vanessa atalanta*) is a well-known colourful butterfly, found in temperate Europe, Asia and North America. The Red Admiral has a 45–50 mm wing span. Two metallic blue spots are on each hindwing. On the underside the forewings are similarly patterned but the hindwings are richly patterned in various shades of brown. The female and male share the same patterning and have wingspans of about 5.5 to 6 cm (about 2 to 2.5 in). The species is resident only in warmer areas, but migrates north in spring, and sometimes again in autumn. This large butterfly is identified by its striking dark brown, red, and black wing pattern. More specifically, the dark wings possess orange bands that cross the fore wings and on the outer edge of the hind wings; white spots on the dorsal fore wings near the front margin; reddish bars on dorsal surface of all four wings The caterpillar feeds on nettles, and the adult drinks from flowering plants like the *Buddleia* and overripe fruit.

In northern Europe, it is one of the last butterflies to be seen before winter sets in, often feeding on the flowers of ivy on sunny days. The Red Admiral is also known to hibernate, re-emerging individuals showing prominently darker colourings than first brood subjects. The butterfly also flies on sunny winter days, especially in southern Europe. In North America, the Red Admiral generally has two broods from March through October. Most of North America must be recolonized each spring by southern migrants, but this species over-winters in south Texas. The Red Admiral is the butterfly featured by Vladimir Nabokov, an amateur lepidoptrist, in his novel "Pale Fire".

Butterflies arriving in Britain in the spring lay eggs on the tips of stinging nettle leaves. The eggs develop into black, spiny caterpillars that live within individual tents, which they construct by spinning together the edges of a nettle leaf with silk. Butterflies also feed on nettle leaves. After about four weeks the caterpillars pupate among the nettle leaves and butterflies emerge about two weeks later. The butterflies are known to hibernate in southern Europe, few survive winter in the British Isles and they are replaced by a new wave of migrants the following spring. Red admirals live in a wide range of habitats, but are most frequently found in wooded countryside, parks, and gardens. The adult butterflies feed at the flowers of many nectar-bearing plants, particularly buddleia, and in the autumn are also attracted to the juices of rotting fruit.

SKIPPER (BUTTERFLY)

Kingdom: Animalia, **Phylum**: Arthropoda, **Class**: Insecta, **Order**: Lepidoptera, (*unranked*): Rhopalocera **Superfamily**: **Hesperioidea, Family**: **Hesperiidae** Latreille, 1809

A skipper or skipper butterfly is a butterfly of the family Hesperiidae. They are named after their quick, darting flight

habits. There are more than 3500 recog-nized species of skippers and they occur worldwide, but with the greatest diversity in the Neotropical regions of Central and South America.

Skippers are considered a type of butterfly. The Order Lepidoptera commonly is separated into two groups, the butterflies and the moths. A lepidopteran generally is considered a butterfly if it belongs either to the superfamily Papilionoidea ("the true butterflies") or the superfamily Hesperioidea ("the skippers"), the latter of which has only one family, Hesperiidae. Some authors also would include as butterflies the members of the superfamily Hedyloidea ("the American butterfly moths").

Although the skippers are usually counted as butterflies, they are somewhat intermediate between the rest of the true butterflies and the remaining Lepidoptera, the moths. There are about 4,000 known species of skippers distributed among 567 genera. Many species of skippers look frustratingly alike. For example, some species in the genera *Erynnis*, *Hesperia*, and *Amblyscirtes* cannot currently be distinguished in the field by experts, the only reliable method of telling them apart involving dissection and exami-nation of the genitalia. Taxonomy of the subfamilies remains unsettled. For example, some authors recognize the "giant skippers" as the subfamily Megathyminae, while others recognize them as a family, or a group of genera within the subfamily Hesperiinae

Subfamilies

There are about 3500 species of skippers. They are now classified in the following subfamilies:

- Coeliadinae – awls, awlets, and policemen (about 75 species)
- Euschemoninae – Regent Skipper (monotypic)

- Eudaminae
- Pyrginae – spread-winged skippers and firetips (including Pyrrhopyginae)
- Heteropterinae – skipperlings (about 150 species)
- Hesperiinae – grass skippers (over 2000 species)
- Megathyminae – giant skippers (about 18 species; doubtfully distinct from Hesperiinae)
- Trapezitinae – Australian skippers (about 60 species)

SWALLOWTAIL BUTTERFLY

Kingdom: Animalia, **Phylum**: Arthropoda, **Class**: Insecta, **Order**: Lepidoptera, **Suborder**: Ditrysia, **Superfamily**: Papilionoidea, **Family**: Papilionidae Latreille, [1802].

Swallowtail butterflies are large, colorful butterflies that form the family Papilionidae. There are over 550 species, and though the majority are tropical, members of the family are found on all continents except Antarc-tica. The family includes the largest butterflies in the world, the birdwing butterflies of Australia (genus *Ornithoptera*). Swallowtails differ from all other butterflies in a number of anatomical traits. Most notably, their caterpillars possess a unique organ behind their heads, called the osmeterium. Normally hidden, this forked structure can be everted when the caterpillar is threatened, or forced out with a gentle squeeze, and emits smelly secretions containing terpenes. The adults are often *tailed* like the forked tail of some swallows, giving the insect its name.

The **Zebra Swallowtail Butterfly** (*Eurytides marcellus*, other authorities list the species under genera in the family including *Iphiclides, Graphium* and *Papilio*) is a swallowtail butterfly found in the eastern United States, north-east Mexico and south-east Canada. Its distinctive black and white-striped pattern is remini-scent of a zebra. The Zebra Swallowtail Butterfly is also the official state butterfly of Tennessee.

RED UNDERWING MOTH

Kingdom: Animalia, **Phylum**: Arthropoda, **Class**: Insecta, **Order**: Lepidoptera, **Superfamily**: Noctuoidea, **Family**: Noctuidae, **Genus**: *Catocala*, **Species**: *Catocala nupta.*

The Red Underwing moth (*Catocala nupta)*, is a moth of Noctuidae family. This is a large (80 mm wingspan) nocturnal European species which, like most noctuids, is drably coloured to aid concealment during the day.

It flies in August and September, and comes freely to both light and sugar. This species and other *Catocala* moths have brightly coloured under-wings, in this case more orange than red. These are not visible at rest, being hidden by the dull fore-wings, but they help the moth avoid predators such as birds if it is disturbed during the day. As the Red underwing moth takes off, the sudden flash of colour may confuse the attacker, and when it lands and immediately closes its wings it may seem to disappear as the colour is "switched off".

It is also thought that the symmetrical patterned orange sections on the rear wings form the illusion of another smaller creature (butterfly), so the attacker will go for the colourful "small

illusive" rear safe region on the main body of this Red underwing moth species. The adult feeds on nectar, the caterpillars eat willow and poplar leaves. Most underwing moths are active at night, and spend the day resting upside down with their wings open against the bark of trees or stumps.

The exposed forewings of underwings provide marvelous camouflage, blending into the background on which the moth rests. Experiments have shown that underwings will select tree barks that increase their degree of camouflage. For example, the white underwing often rests on the bark of white birch trees while the blackish wayward nymph typically rests on dark tree trunks or blackened stumps. When a bird or other potential predator disturbs a camouflaged underwing moth, the sudden, brilliant flash of color when the moth takes off may momentarily startle or confuse the predator, allowing the moth an extra moment to escape.

The bright hindwing colours may also deflect an attack away from the moth's body, resulting in wing damage but not death. During the summer, the female underwing deposits her eggs one at a time, or in small clusters, into tree bark crevices. Underwings commonly deposit their eggs on willows, poplars, walnuts, hickories, apple, and cherry trees. The hardy eggs are able to withstand extremely cold and dry weather and do not hatch until the following spring. As the leaf buds of the underwing's host plants start to open, the eggs hatch into caterpillars, which then feed on the growing leaves.

VICEROY BUTTERFLY

Kingdom: Animalia, **Phylum**: Arthropoda, **Class**: Insecta, **Order**: Lepidoptera, (**unranked**): Rhopalocera, **Superfamily**: Papilionoidea, **Family**: Nymphalidae, **Subfamily**: Limenitidinae, **Genus**: *Limenitis*, **Species**: *Limenitis archippus.*

The Viceroy Butterfly (*Limenitis archippus*) is a North American butterfly with a range from the Northwest Territories along the eastern edges of the Cascade Range and Sierra Nevada mountains, southwards into central Mexico. Its wings feature an orange and black pattern, and over most of its range it is a Müllerian mimic with the Monarch butterfly (*Danaus plexippus*).

Wingspan is between 53 and 81 mm. The Monarch is bigger and doesn't have the postmedian black line that runs across the veins on the hindwing. In Florida, Georgia, and the South-west, Viceroys share the pattern of the Queen Butterfly (*D. gilippus*) and in Mexico they share the pattern of the Soldier Butterfly (*D. eresimus*). In all three areas, the local *Danaus* population mimic the coloration of the Viceroy species. It was originally believed that the Viceroy was a Batesian mimic of the three other species, in that it was presumably edible or only mildly unpalatable to predators, but this has since proven not to be true.

Life Cycle

The eggs of the viceroy are laid at the tips of oak, willow, birch, or linden leaves and are spherical in shape. The surface of the egg is covered with bristles and pitted with six-sided cells. The larva, or caterpillar, feeds on the leaf tip on which it was hatched and attaches bits of partially chewed leaf to the leaf midrib by strands of silk. When winter approaches, the caterpillar detaches the uneaten portions of the leaf from the midrib, glues the rib of the leaf to the stem with a stout silk thread, rolls itself up in the detached leaf to form a hibernaculum, or winter quarters, and reattaches the leaf to the rib. The caterpillar feeds on trees in the willow family

Salicaceae, including willows (*Salix*), and poplars and cottonwoods (*Populus*).

The caterpillars sequester the salicylic acid in their bodies, which makes them bitter, and upsets predators' stomachs. As further protection, the caterpillars, as well as their chrysalis stage, resemble bird droppings. Adults are strictly diurnal, they fly preferentially in the late morning and early afternoon. The Viceroy was named the state butterfly of Kentucky in 1990. The viceroy is unpalatable to insectivorous animals. Adults bear a remarkable color resemblance to the monarch butterfly, and the two species are *Mullerian mimics*.

Caddisflies

Kingdom: Animalia, **Phylum**: Arthropoda, **Class**: Insecta, **Superorder**: Amphiesmenoptera, **Order**: Trichoptera.

Caddisfly (*Phryganea grandis*) is Britain's largest caddisfly and is also called the large red sedge. There are about 190 other caddisfly species in Britain.

The caddisflies are an order, Trichoptera, of insects with approximately 12,000 described species. Also called sedge-flies or rail-flies, they are small moth-like insects having two pairs of hairy memb-ranous wings. They are closely related to Lepi-doptera (moths and butter-flies) which have scales on their wings, and the two orders together form the superorder Amphies-menoptera. Caddisflies have aquatic larvae and are found in a wide variety of habitats such as streams, rivers, lakes, ponds, spring seeps, and temporary waters (vernal pools). The name "Trichoptera" comes from Greek: *trich*, "hair" + *ptera*, "wings". Caddisfly eggs are deposited in green, gelatinous masses and are usually attached to water plants. The larvae, called caddisworms, caseworms, or cadbait (from their wide use for bait by anglers), are aquatic. The caddisworm is elongate with a small, hard head and a soft thorax and abdomen. For protection the larva lives in a characteristic tubular silk case covered with sand grains, bits of shell, twigs, or leaves. Most caddisworms are herbivorous; several species, however, also eat small aquatic animals, which they catch in nets spun from secreted silk and anchored in running water. After several months in the larval stage, the caddisworm closes the mouth of the case and pupates. At the end of the pupal stage the pupa rises to the water's surface, and the adult quickly emerges and flies away.

Life Cycle

A caddisfly, sometimes called a sedge fly, is an aquatic cousin of moths and butterflies. A caddisfly goes through a complete life cycle of egg, larva, pupa and adult phase. The life cycle of a caddisfly typically takes six months to 2 years to complete.

Egg: The females of both caddisfly species lay their eggs on plants overhanging a body of water or on rocks just beneath the water's surface. The eggs sink to the bottom and hatch into larvae in about 10 to 12 days.

Larva: The caddisfly larva phase typically lasts for six to 10 months. Mature larvae wrap themselves in a silk netting to protect themselves from predators while they pupate.

Pupa: A cadisfly pupa spends about two to five weeks maturing in its cocoon. A mature caddisfly pupa crawls or swims to the water's surface, where it sheds its transparent pupal membrane and emerges as an oxygen-breathing adult.

Adult: Adult caddisflies live for two weeks to two months. The adult females participate in mating swarms, and then fly off to lay their eggs.

Habitat and Food

Most caddisfly larvae and pupae (PYU-pee), the life stage between larva and adult, are found in freshwater, but there are a few species that live on land or in the sea. The larvae of freshwater species usually live in cold clean flowing waters, but some species prefer warmer slower waters. They are very particular about water temperature and speed, dissolved minerals and pollutants, as well as the amount of sunlight. Several species can live together in a stream or river because each occupies habitats within the water that do not overlap. Predatory species wander about freely in the environment hunting for food animals, but many plant-feeding species live in protective cases built from pebbles, sand, or bits of vegetation from the bottom and held together with silk. A marine species from Australia and New Zealand spends part of its larval life eating tissues inside the body cavity of a living sea star (starfish). Later it leaves its host to build a case from seaweed. The adults are usually active at night. They spend their days hiding in moist, cool habitats and are often found on vegetation growing along river banks. Most caddisfly species eat algae, decaying plants and organic animal matter, there are a few species that are predators.

Chalcid Wasp

Kingdom: Animalia, **Phylum**: Arthropoda, **Class**: Insecta, **Order**: Hymenoptera, **Suborder**: Apocrita, **Superfamily**: **Chalcidoidea**

Chalcid wasps (superfamily Chalcidoidea) belong to the insect order Hymenoptera, and are one of the largest groups within the order, with some 22,000 known species, and an estimated total diversity of anywhere from 60,000 to more than 500,000 species, meaning the vast majority have yet to be discovered and described. Most of the species are parasitoids of other insects, attacking the egg or larval stage of their host, though many other life cycles are known.

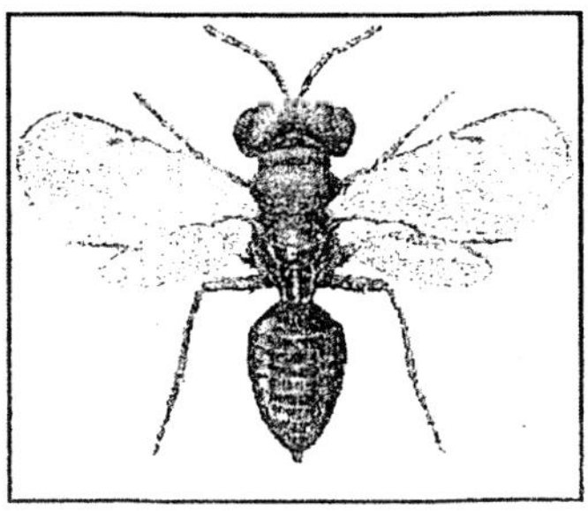

These hosts are to be found in at least 12 different insect orders including Lepidoptera (butterflies and moths), Diptera (true flies), Coleoptera (beetles), Hemiptera (true bugs), and other Hymenoptera, as well as two orders of Arachnida, and even one family of nematodes. A few species are phytophagous and the larvae feed inside seeds, stems, and galls.

Generally beneficial to humans as a group, chalcids keep various crop pests under control, and many species have been imported to control insect pests. They can be found almost everywhere, particularly on flowers, foliage, and in leaf litter, but are often overlooked due to their very small size, which includes the smallest of all known insects.Chalcids are tiny, dark-coloured wasps, often metallic blue or green with complex sculpturing on the body. They are also recognized by the characteristic reduced wing venation, similar to that seen in other superfamilies of parasitic wasps. They have a significant part to play in ecosystems and their true importance must not be underestimated.

Chalcidoidea is a superfamily of Hymenoptera, whose family constituency is in constant flux, as new hypotheses of relationships are constantly being proposed and rejected; with the advent of molecular systematics, it seems that the future will see further revisions of the classification in use today. There are nineteen extant families recognized at present:

- Agaonidae, Walker, 1846
- Aphelinidae, Thomson, 1876
- Chalcididae, Latreille, 1817
- Encyrtidae, Walker, 1837
- Eucharitidae, Latreille, 1809
- Eulophidae, Westwood, 1829 (including Elasmidae)
- Eupelmidae, Walker, 1833
- Eurytomidae, Walker, 1832
- Leucospidae, Fabricius, 1775
- Mymaridae, Haliday, 1833
- Ormyridae, Förster, 1856
- Perilampidae, Latreille, 1809
- Pteromalidae, Dalman, 1820
- Rotoitidae, Bouek & Noyes, 1987
- Signiphoridae, Ashmead, 1880
- Tanaostigmatidae, Howard, 1890
- Tetracampidae, Förster, 1856
- Torymidae, Walker, 1833
- Trichogrammatidae, Haliday, 1851

There is also one extinct family, **Khutelchalcididae.**

Life Cycle

"All chalcidids are parasitic. Most attack pupae of Lepidoptera and Diptera, but some parasitize other Hymenoptera or beetles. Parasites of Lepidoptera usually attack young pupae, while those of Diptera attack mature larvae.

Habitat

On the underside of a Bramble leaf suffering from Violet Bramble-rust in garden.

Economic Importance and Biological Diversity

Although often very beautiful because of their unusual structures and striking colours or metallic hues, chalcids are very tiny and unobtrusive, not biting or stinging humans, living in their domiciles, or carting off their food. Because of this they remain largely unnoticed. They are unknown to all but the specialist and are under-appreciated for their role in the environment and benefit to man. About 80 chalcid species are known to be pests of agriculture (mostly seed-feeders in the families Eurytomidae and Torymidae) and some chalcids are considered harmful because they are hyperparasitoids, but most are economically and environmentally beneficial. The large majority of chalcid species are primary parasitoids of other insects and arachnids and as such they are important participants in nature's own control system for regulating arthropod populations. In addition to the largely unappreciated role of most species in helping to control what might otherwise be pest species, over 800 chalcid species have been associated with targeted biological control programs.

This represents about two-thirds of all biocontrol programs involving Hymenoptera, and about one-third of all biocontrol programs in which partial or complete economic control of an insect pest was achieved. Members are known to attack hosts in about 340 families of 15 insect orders (Blattaria, Coleoptera,

Diptera, Hemiptera, Homoptera, Hymenoptera, Lepidoptera, Mantodea, Neuroptera, Odonata, Orthoptera, Psocoptera, Siphonaptera, Strepsiptera and Thysanoptera), as well as egg sacs of spiders (Araneae), ticks and gall-forming mites (Acari), cocoons of pseudoscorpions (Pseudoscorpiones), and gall-forming Anguinidae (Nematoda). In addition, phytophagous chalcids are known from six families. Agaoninae (Agaonidae) are exclusively phytophagous within the ovarioles of figs (*Ficus*), but are beneficial as the obligate pollinators of figs. Some members of five other families (Eulophidae, Eurytomidae, Pteromalidae, Tanaostigmatidae and Torymidae) are seed-feeders or gall-formers on plants, though for many chalcids reared from galls it is not known whether they are primary gall-formers, or inquilines or parasitoids in the galls.

Possibly as a consequence of their host diversity, chalcids demonstrate all but 2 of 15 feeding types that have been defined for insects, including parasitism of all host life stages from egg to adult, as internal or external parasitoids, as primary or hyperparasitoids, and with their eggs laid in, on or away from the host. Some chalcids of the family Aphelinidae even parasitize the opposite sex of their own species. Like other Hymenoptera, most chalcids have a haploid-diploid mechanism of sex determination in which fertilized (diploid) eggs normally develop into females and unfertilized (haploid) eggs normally develop into males (arrhentokous development).

However, males are unknown or are very rare for some species and in these species females produce females asexually (thelytokous development). This latter condition appears often to be the result of the presence of maternally inherited cytoplasmic microorganisms, bacteria in the genus Wolbachia. Heat or such antibiotics as tetracycline can kill the bacterium and turn a thelytokous strain into an arrhentokous strain. It has also been shown that reproductive isolation between closely related species is at least sometimes the result of the presence of Wolbachia or of the species possessing different bacterial 'types'. Viable hybrid offspring can be produced if the species are 'cured' of their endosymbiotic bacterium.

Chigoe Flea

Kingdom: Animalia, **Phylum**: Arthropoda, **Class**: Insecta, **Order**: Siphonaptera, **Family**: Hectopsyllidae, **Genus**: *Tunga* Species: *Tunga penetrans.*

The chigoe flea or jigger (*Tunga penetrans*) is a parasitic arthropod found in tropical climates, especially South and Central America and the West Indies, not to be confused with the larval form of Trombiculidae (called chiggers) found in more temperate climates. In Brazil, the parasite is referred to as bicho de pé. At 1 mm long, the chigoe flea is the smallest known flea. Breeding female chigoes burrow into exposed skin on the feet of mammals and remain there for two weeks while developing eggs, sometimes causing intense irritation. After this point, the skin lesion looks like a 5 to 10 mm blister with a central black dot, which are the flea's exposed hind legs, respiratory spiracles and reproductive organs. If the flea is left within the skin, infection and/or other dangerous complications can occur, although they are relatively rare.

The parasitic flea lives in soil and sand, and feeds intermittently on warm-blooded hosts such as humans, cattle, sheep, dogs, mice, and other animals. In order to reproduce, the female flea burrows head-first into the hosts' skin, often leaving the caudal tip of its abdomen visible through an orifice in a skin lesion. This orifice allows the chigoe flea to breathe and defecate while feeding on blood vessels in the cutaneous and subcutaneous dermal layer. In the next two weeks, its abdomen swells with up to several dozen eggs which it releases through the caudal orifice to fall to the ground when ready to hatch. The flea then dies and is sloughed off with the host's skin. Within the next three to four days, the eggs hatch and mature into adult fleas within three to four weeks. Infections are almost always on the foot of the host.

During the first day or two of infection the host may feel an itching or irritation which then passes as the area around the flea calluses and becomes insensitive. As the flea's abdomen swells with eggs later in the cycle, the pressure from the swelling may press neighbouring nerves or blood vessels. Depending on the exact site, this can cause sensations ranging from mild irritation to serious discomfort. It has been written that the explanation for the preponderance of tungiasis lesions on the toes is because chigoe flea is a poor jumper, attaining only a high of 20 cm.

But this idea is not correct for many reasons. First of all, while chigoe flea can jump up only 20 cm, its infamous and bigger cousin Flea which is the vector of bubonic plague and well known for its jumping ability, can jump up only 18 cm; the chigoe flea should be considered a champion jumper, not a poor one. And the jumping ability cannot explain why hands are the second most affected body part. The occurrence of tungiasis lesions on the toes, between them, on the soles can easily explained by the fact that most of the victims are poor and walk barefoot. Because the front part of the feet, which are the toes hit the ground first, the fleas have more opportunity to latch on there. They also get trapped between the toes, or under the soles when the victims stand still. More lesions on the hands are seen in children because they play in the sand or dusty ground. Other common ectopic sites of infestation are the elbows, buttocks and the genital area (gluteal area in scientific publications) and this can be explained by the habit of sitting on bare ground while leaning back on the elbows.

Life Cycle and Habitat

Fleas are holometabolous insects, going through the four life cycle stages of egg, larva, pupa, and imago (adult). The flea life cycle begins when the female lays after feeding. Adult fleas must feed on blood before they can become capable of reproduction. Eggs are laid in batches of up to 20 or so, usually on the host itself, which means that the eggs can easily roll onto the ground. Because of this, areas where the host rests and sleeps become one of the primary habitats of eggs and developing fleas. The eggs take around two days to two weeks to hatch._Fleas pass through a complete life cycle consisting of egg, larva, pupa and adult.

A typical flea population consists of 50% eggs, 35% larvae, 10% pupae and 5% adults. Completion of the life cycle from egg to adult varies from two weeks to eight months depending on the temperature, humidity, food, and species. Normally after a blood meal, the female flea lays about 45 to 50 eggs per day up to 600 in a lifetime usually on the host (dogs, cats, rats, rabbits, mice, squirrels, chipmunks, raccoons, opossums, foxes, chickens, humans, etc.). Eggs loosely laid in the hair coat, drop out most anywhere especially where the host rests, sleeps or nests (rugs, carpets, upholstered furniture, cat or dog boxes, kennels, sand boxes, etc.).

Eggs hatch in two days to two weeks into larvae found indoors in floor cracks & crevices, along baseboards, under rug edges and in furniture or beds. Outdoor development occurs in sandy gravel soils (moist sand boxes, dirt crawlspace under the house, under shrubs, etc.) where the pet may rest or sleep. Sand and gravel are very suitable for larval development which is the reason fleas are erroneously called "sand fleas."

Larvae are blind, avoid light, pass through three larval instars and take a week to several months to develop. Their food consists of digested blood from adult flea feces, dead skin, hair, feathers, and other organic debris. (Larvae do not suck blood.) Pupa mature to adulthood within a silken cocoon woven by the larva to which pet hair, carpet fiber, dust, grass cuttings, and other debris adheres. In about five to fourteen days, adult fleas emerge or may remain resting in the cocoon until the detection of vibration (pet and people movement), pressure (host animal lying down on them), heat, noise, or carbon dioxide (meaning a potential blood source is near). Most fleas overwinter in the larval or pupal stage with survival and growth best during warm, moist winters and spring.

Adult fleas cannot survive or lay eggs without a blood meal, but may live for one year without feeding. There is often a desperate need for flea control after a family has returned from a long vacation. The house has been empty with no cat or dog around for fleas to feed on. When the family and pets are gone, flea eggs hatch and larvae pupate. The adult fleas fully developed inside the pupal cocoon remains in a kind of "limbo" for a long time until a blood source is near. The family returning from vacation is immediately attacked by waiting hungry hordes of fleas. (In just 30 days, 10 female fleas under ideal conditions can multiply to over a quarter million different life stages.)

Newly emerged adult fleas live only about one week if a blood meal is not obtained. However, completely developed adult fleas can live for several months without eating, so long as they do not emerge from their puparia. Optimum temperatures for the flea's life cycle are 70°F to 85°F and optimum humidity is 70 per cent.

Cicada

Kingdom: Animalia, **Phylum**: Arthropoda, **Class**: Insecta, **Order**: Hemiptera, **Suborder**: Auchenorrhyncha, **Infraorder**: Cicadomorpha, **Superfamily**: **Cicadoidea, Family**: **Cicadidae** Westwood, 1840.

A cicada is an insect of the order Hemiptera, suborder Auchenorrhyncha, in the superfamily Cicadoidea, with large eyes wide apart on the head and usually transparent, well-veined wings. There are about 2,500 species of cicada around the world, and many remain unclassified. Cicadas live in temperate to tropical climates where they are among the most widely recognized of all insects, mainly due to their large size and remarkable acoustic talents. Cicadas are often colloquially called locusts, although they are unrelated to true locusts, which are a kind of grasshopper. They are also known as "jar flies" and, regionally, as July flies in the Southeastern United States, and as "heat bugs" in Canada and the mid-West.

Cicadas are related to leafhoppers and spittlebugs. In parts of the southern Appalachian Mountains in the United States, they are known as "dry flies" because of the dry shell that they leave behind. Cicadas are benign to humans in normal circumstances and do not bite or sting in a true sense, but may in fact sting after mistaking a person's arm or other part of their body as a tree or plant limb and attempt to feed. Cicadas have a long proboscis under their head that they use for feeding on tree sap, and if they attempt to.inject it into a person's body it can be painful, but is in no other way harmful. This sting is not a defensive reaction and should not be mistaken for aggression; it is extremely uncommon, and usually only happens when they are allowed to rest on a person's body for an extended amount of time. Cicadas can cause damage to several cultivated crops, shrubs, and trees, mainly in the form of scarring left on tree branches while the females lay their eggs deep in branches. Many people around the world regularly eat cicadas; the female is prized, as it is meatier.

Cicadas have been (or are still) eaten in Ancient Greece, China, Malaysia, Burma, Latin America, and the Congo. Shells of cicadas are employed in the traditional medicines of China. The name is a direct derivation of the Latin *cicada,* meaning "buzzer". In classical Greek, it was called a *tettix,* and in modern Greek *tzitzikas*—both names being onomatopoeic.

After mating, the female cuts slits into the bark of a twig, and into these she deposits her eggs. She may do so repeatedly, until she has laid several hundred eggs. When the eggs hatch, the newborn nymphs drop to the ground, where they burrow. Most cicadas go through a life cycle that lasts from two to five years. Some species have much longer life cycles, such as the North American genus, *Magicicada,* which has a number of distinct "broods" that go through either a 17-year or, in the South of the USA, a 13-year life cycle. These long life cycles perhaps developed as a response to predators such as the cicada killer wasp and praying mantis. A predator with a shorter life cycle of at least 2 years could not reliably prey upon the cicadas. Cicadas live underground as nymphs for most of their lives, at depths ranging from about 30 cm (1 ft) down to 2.5 m (about 8½ ft).

The nymphs feed on root juice and have strong front legs for digging. In the final nymphal instar, they construct an exit tunnel to the surface and emerge. They then molt (shed their skins), on a nearby plant for the last time and emerge as adults. The abandoned skins remain, still clinging to the bark of trees.

Male cicadas have loud noisemakers called "tymbals" on the sides of the abdominal base. Their "singing" is not the stridulation (where one structure is rubbed against another) of many other familiar sound-producing insects like crickets: the tymbals are regions of the exoskeleton that are modified to form a complex membrane with thin, membranous portions and thickened "ribs". Contracting the internal tymbal muscles produces a clicking sound as the tymbals buckle inwards.

Life Cycle

Gradual metamorphosis. Cicadas, also known as 'locusts', appear as adults in May and June, and crawl up into host trees to mate. After mating, the female cicada uses a knife-like organ to slit or puncture twigs of woody plants. She then lays eggs in the slits. In about 6 weeks the eggs hatch and the nymphs drop to the ground. They then dig into the soil and begin feeding by sucking sap from the tree roots. After approximately 3 years, the nymphs come out of the ground. They crawl up tree trunks, posts, or other objects; shed their last exoskeleton and emerge as adult, winged cicada. These adults live about one month. During this time they mate, and each female lays from 400 to 600 eggs.

Habitat

Around treas and other natural habitat.

Food

Adults do not eat. Nymphs feeds on juices from roots.

Cockroach

Kingdom: Animalia, **Phylum**: Arthropoda, **Class**: Insecta, **Subclass**: Pterygota, **Infraclass**: Neoptera, **Superorder**: Dictyoptera, **Order**: Blattaria.

Cockroaches are insects of the order *Blattaria.* The name derives from the Latinized Greek name for the insect (Doric Greek: *blátta;* Ionic and Attic Greek: *blátte;* Latin: *Blatta*). There are about 4,500 species of cockroach, of which 30 species are associated with human habitations and about four species are well-known as pests. Among the best-known pest species are the American cockroach, *Periplaneta americana,* which is about 30 millimetres (1.2 in) long, the German cockroach *(Blattella germanica),* about 15 millimetres (0.59 in) long, the Asian cockroach *(Blattella asahinai),* also about 15 millimetres (0.59 in) in length, and the Oriental cockroach *(Blatta orientalis),* about 25 millimetres (0.98 in).

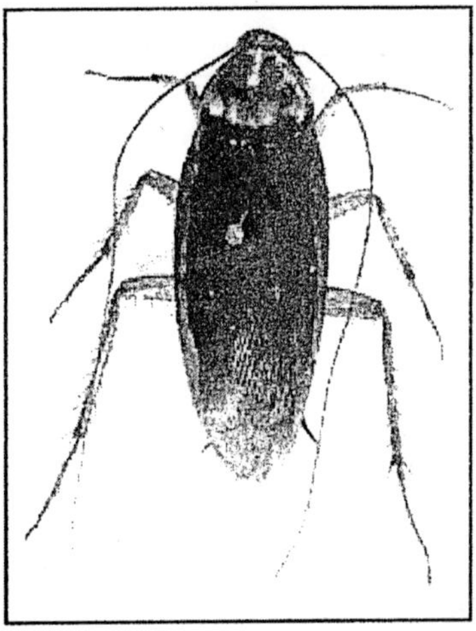

Tropical cockroaches are often much bigger, and extinct cockroach relatives and 'roachoids' such as the Carboniferous *Archimylacris* and the Permian *Apthoroblattina* were not as large as the biggest modern species.

Species of Cockroaches

- *Blattella germanica*, German cockroach
- *Blaptica dubia*, South American/Peruvian Dubia cockroach
- *Blatta orientalis*, Oriental cockroach
- *Blattella asahinai*, Asian cockroach
- *Blaberus craniifer*, true death's head cockroach
- *Blaberus discoidalis*, discoid cockroach or false death's head
- *Eurycotis floridana*, Florida woods cockroach
- *Gromphadorhina portentosa*, Madagascar hissing cockroach
- *laxta granicollis*, Bark cockroach
- *Parcoblatta pennsylvanica*, Pennsylvania woods cockroach
- *Periplaneta americana*, American cockroach
- *Periplaneta australasiae*, Australian cockroach
- *Periplaneta brunnea*, black Mississippi cockroach
- *Periplaneta fuliginosa*, smokybrown cockroach
- *Pycnoscelus surinamensis*, Surinam cockroach
- *Supella longipalpa*, brown-banded cockroach.

Cockroaches in general are flat and oval in outline. The head points downward and is protected by the enlarged flanges of the thorax. In many other respects they are similar in morphology to grasshoppers, crickets, katydids, stick insects, and mantids, their closest relatives. Cockroaches may be winged or wingless, but even winged species do not always fly. Adults range from 1 mm (0.04 in) to more than 9 cm (3.6 in) in length. They are sensitive to light; most species prefer darkness, being la:gely nocturnal. Their long, hairlike, highly sensitive antennae and sensory bristles enable them to detect tiny amounts of food and moisture. The cerci, sensory structures extending from the rear of the abdomen, can sense minute air movements, enabling the cockroach to rapidly detect and flee from potential danger. Most cockroaches can run very rapidly and are difficult to catch due to their soft, slippery outer skin, called the cuticle. They can hide in very narrow crevices. The oily cuticle also protects them from dehydration. Individual species may be restricted to very specific habitats such as leaf litter, bromeliads, the splash zone of waterfalls, or bat caves. Some species are brightly coloured, defying the stereotype of the drab, brown or black household roach. Some species, including the Madeira cockroach, can produce sound.

Life Cycle and Reproductive Behaviour

Courtship patterns vary with the species. When the American cockroach is ready to mate, the female produces a chemical odoue, or pheromone, that attracts males. The males flap their wings and probe for females with their abdomens when they sense the pheromone; eventually, they back into a female and mating ensues. Other species have more elaborate mating displays, including hissing noises, bobbing and waving of the abdomen, or nibbling. The males of one African species form dominance hierarchies, and the females preferentially mate with the dominant male.

The fertilized eggs that result from mating are cemented together by the female in a sausage-shaped egg case known as the ootheca. The American cockroach simply deposits the ootheca in a protected place and abandons it. German cockroaches keep the ootheca extended from their egg-laying organ, known as the ovipositor, depositing it only when the eggs are ready to hatch. An ootheca contains from 16 to 32 eggs, depending on the species. Female ootheca-tending cockroaches sometimes display maternal care. The nymphs that emerge from the eggs often remain around their mother for several days. Nymphs undergo gradual metamorphosis-that is, they grow and mature in stages, each separated by a molt of the exoskeleton. Each successive stage, or instar, comes to resemble the adult insect more and more. The final molt gives rise to a winged, sexually mature individual.

Habitat

Cockroaches live anywhere! You can find them all over the world. They live indoors, like in kitchens, classrooms, or any warm snug place. They live outdoors, like in forests close to water, and the tropics. Cockroaches love wallpaper, cloth, ink, and cardboard boxes. They like anything that might have startch or protein. They will hitchike by hiding in boxes. Then they get loaded onto boats and wagons to find warm spots to live.

Importance

Cockroaches play an important role in ecology. They help decompose forest litter and animal fecal matter and are, in turn, food for many other animals, including scavengers of dead cockroaches, predators, and egg parasites. They are generalist feeders, capable of digesting a wide range of substances due to the variety of bacteria and protozoa in their digestive systems. The success of their survival strategies is proven by the longevity of the group and their amazing diversity. They are therefore an important part of the food web. Cockroaches have earned a bad reputation not only because they feed on human food stores and garbage but because they foul their surroundings, leave behind a repugnant odoue, and are extremely difficult to eradicate. About one per cent of cockroach species are considered pests to humans and these are basically nuisance pests. In laboratory experiments, household species are capable of contaminating food and other objects with human disease organisms, but they have not been implicated in actual disease outbreaks. In fact, cockroaches spend a great deal of time cleaning themselves. Cockroaches have been the target of many insecticides over the years but they have developed resistance to several of them.

Once the cuticle is abraded, the roaches die of dehydration. Also effective is boric acid powder, which is both abrasive and poisonous to cockroaches. The best way to prevent cockroaches from multiplying is to keep a clean house and block their access to water, which they need to survive. Attempts to use pheromones as sex lures or to sterilize male cockroaches have thus far not proved practical on a large scale. Sprinkling abrasives such as diatomaceous earth to penetrate their protective cuticles may work in individual households as a nonpoisonous alternative.

Control

When you spot cockroaches at home, try figuring out where they are coming from. But when you actually get down to doing it, you will discover that this is not all that easy. Roaches normally live in groups. So check for possible cracks and holes in your home, which they might have converted into a living zone. Once you have spotted it, you can clean up the place. You can also kill the cockroaches if you want. There are many ways to do this. For example, you can use an insecticide spray, cockroach bait or a roach trap, etc. You can use a liquid concentrate for getting rid of cockroaches. Such concentrates are basically poison or deterrent chemical that is thinned with water.

It can be sprayed, wiped or mopped onto just about any type of surface. It will remain effective for a week or so preventing the re-infestation of the roaches. A very simple way to kill cockroaches is by using a solution of soap and water. Make a light concentrate and spray this on the body of the roach. Since roaches breathe through their skin, the forming of a thin soap film will block their breathing pores on the skin. This way, the roach will suffocate to death or near death.

The best way to prevent re-infestation of cockroaches in the house is by keeping it clean. Take care to regularly clean your house, especially all nooks and corners if you do not want the roaches to resurface. Get help from a professional. Incase all the above mentioned roach control tips do not work, then call a pest-control expert. These people are licensed to use stronger chemicals and they can do so while keeping your family safe.

Cricket

Kingdom: Animalia, **Phylum**: Arthropoda, **Class**: Insecta, **Order**: Orthoptera, **Superfamily**: Gryiloidea, **Family**: **Gryllidae** Bolívar, 1878.

Crickets, family Gryllidae (also known as "true crickets"), are insects somewhat related to grasshoppers, and more closely related to katydids or bush crickets (family Tettigoniidae). They have somewhat flattened bodies and long antennae. There are about 900 species of crickets. They tend to be nocturnal and are often confused with grasshoppers because they have a similar body structure including jumping hind legs. Crickets are harmless to humans. Crickets are omnivorous scavengers who feed on organic materials, as well as decaying plant material, fungi, and some seedling plants.

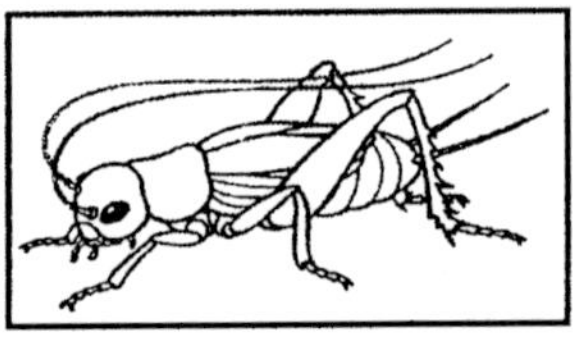

Crickets eat their own dead when there are no other sources of food available, and exhibit predatorial behaviour upon weakened, crippled crickets. Crickets are known for their chirp (which only male crickets can do; male wings have ridges that act like a "comb and file" instrument). They chirp by rubbing their wings or legs over each other, and the song is species-specific. Crickets have relatively powerful jaws, and have been known to bite humans, mostly without breaking the skin. The bite can, however, be painful when inflicted on sensitive skin such as the webbing between fingers.

Crickets have long antennae and hind legs adapted for jumping; organs for hearing are located on their front legs. Solitary by day, crickets remain in crevices, under rocks, or in shallow burrows dug in the soil, emerging at night to feed on plants. During breeding season, the male cricket attracts a female with his call, sometimes driving off other males that intrude on his territory.

The call is distinctive in each species. The snowy tree cricket varies its chirping according to the air temperature. The temperature in Fahrenheit degrees can be estimated simply by adding 40 to the number of its chirps in 15 seconds. In most crickets, after mating the female uses her long, spearlike ovipositor to insert eggs into the soil or plant stems. The young, called nymphs, resemble the adults.

They reach full size after 6 to 12 molts during which they shed their outer covering; as adults, they usually live six to eight weeks.

Life Cycle

Crickets mate in late summer and lay their eggs in the fall. The eggs hatch in the spring and have been estimated to number as high as 200 per fertile female. Subspecies *Acheta Domestica* however lays eggs almost continually, with the females capable of laying at least twice a month. Female crickets have a long needlelike egg-laying organ called an ovipositor. Nymph crickets are very small when they first emerge from their eggs, measuring only about 1/8 inch in length. Crickets usually go through eight to 10 instars before reaching adulthood, which takes about two to three months. Crickets start to develop wings at about 1 month of age. Adult crickets usually measure about 1 inch in length. Adult crickets have bodies with three distinct segments: the head, the thorax and the abdomen. They also have three pairs of legs and two antennae. In most species of cricket, the adults have wings. Adult crickets are usually omnivores, eating insects and plant matter. Crickets usually live for about two months.

Habitat

The FOSS cricket habitat has three chambers. One contains soil that is kept moist. This is where the females will lay eggs.

A second chamber contains dry sand. Food should be placed in this area so that it will not mold. The central area is the exercise yard with structures for climbing and hiding. Crickets need paper, sand, or soil to get around because their feet are not adapted for holding onto smooth surfaces. A large or complex cricket culture container will allow the crickets to display preference for variables such as moisture, temperature, and structure. Crickets prefer a hot, dry environment. If they are kept in a humid environment, they can develop a fungus, so lots of ventilation is needed. They can go into a moist environment to eat or lay eggs, but they must be able to retreat to dry ground. If you like, you can train a lamp on the central part of the habitat—the crickets will congregate there to bask. And you will probably get a song tossed in as part of the bargain.

Control

The first step in cricket control is to check around the outside of your home and caulk or repair cracks and other spaces you find that may allow crickets to get inside. Look carefully around your home's foundation, doors and ground-level windows. You can discourage crickets from entering your home by keeping nearby weeds and grass cut short and by removing piles or stacks of wood, brush, bricks and similar objects that are close to your house. This minimizes their hiding places and reduces the number of crickets that get inside.

If they get inside, don't reach for the aerosol insecticide. You can easily catch crickets by placing sticky glue boards (available at hardware stores, usually for mouse control) in corners of the room. Within a day or two, your crickets will be caught in the glue board. If you've got a chirping cricket inside the house, pour a small amount of cornmeal in the center of a glue board and place it near where you hear the chirping sound. Within a day or two, you should have gotten your cricket.

Dobsonfly

Kingdom: Animalia, **Phylum**: Arthropoda, **Class**: Insecta, **Order**: Megaloptera, **Family**: Corydalidae, **Subfamily**: **Corydalinae**

The Dobsonfly, also known as "The King Bug" for its kingly features and intimidating tusks, is an insect of the subfamily Corydalinae, part of the megalopteran family Corydalidae. The most well-known species is *Corydalus cornutus*, the eastern dobsonfly. This is a long, dark-coloured insect found in North and Central America. Their closest relatives are the fishflies. Both male and female dobsonflies can reach lengths up to five inches (12.5 cm), measured from the tips of their pincers to the tips of their four wings. Their wingspans can be twice as long as their body length. The wings are densely lined with intersecting veins. When not in use, the wings are folded along the length of their walkingstick-like bodies.

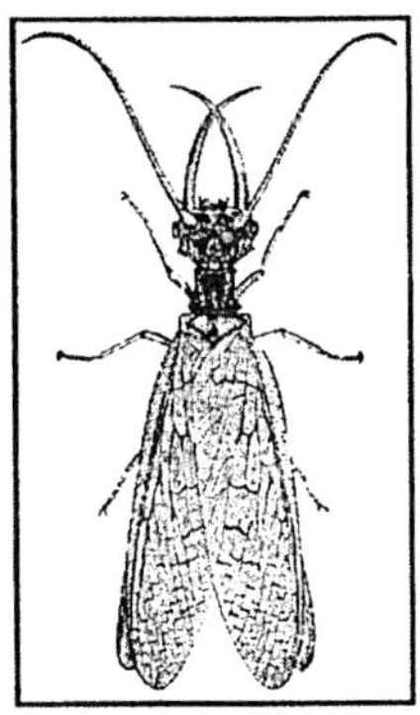

Dobsonflies have segmented antennae similar to ants and wasps. Though both male and female dobsonflies have sharp mandibles, those of an adult male dobsonfly are actually so big – up to 1 inch (25 mm) – that they are unable to harm humans, as they have such poor leverage that they are incapable of breaking the skin. Their mandibles are used exclusively during mating, where males show them off and grasp the females during copulation. Female dobsonflies, however, retain the short, powerful pincers they had as larvae, so they can inflict painful bites, which can draw blood. Not withstanding the males' inability to inflict harm, when threatened both sexes will raise their heads and spread their jaws menacingly.

They are not poisonous, but possess an irritating, foul-smelling anal spray as a last-ditch defense. Dobsonflies spend most of their life in the larval stage, during which they are called hellgrammites, and are familiar to anglers who like to use the large larvae as bait. Hellgrammites live under rocks at the bottoms of lakes, streams and rivers, and prey on other insect larvae with the short sharp pincers on their heads, with which they can also inflict painful bites on humans. The larvae reach to 2" to 3" in length, with gills all along the sides of their segmented bodies that allow them to breathe underwater. Their gills, primitive nervous systems, and hard, segmented bodies mean that once hooked, hellgrammites can survive for long periods wiggling underwater. In the Appalachians, children catch these for fishing bait as a test of courage, working their fingers into the mud under rocks until a hellgrammite bites, then grabbing the creature with the other hand and throwing it into a jar.

A less painful way to catch them is to stretch a mesh net across a narrow point in the stream, forcing the net to the bottom, then have others upstream turn over rocks and disturb the creek bottom, forcing the hellgrammites into the streamflow so they are carried into the net, whence they can be plucked and put into bait buckets. After a few years of living and growing underwater, the larvae crawl out onto land and pupate. They stay in their cocoons over the winter and emerge only to mate. Upon emerging, they live for only seven days. While not generally believed to eat during their adult stage, some captive female specimens have been observed with their heads burrowed into blackberries.

Adults can generally be found from late spring into the middle of summer, preferring to remain near bodies of water, particularly the ones where they grew up. Once they emerge as adults they mate, deposit their eggs near the water (often on overhanging vegetation), then die. They are primarily nocturnal, and like most aquatic insects, are commonly attracted to bright lights. The giant dobsonfly is the largest United States aquatic insect. It occurs in the eastern states. The brownish-green adult of this species is about 7.6 cm (about 3 in) long and has a wingspread of nearly 13 cm (nearly 5 in); the male has slender mandibles that can be up to 2.5 cm (1 in) long.

Life Cycle

Immatures develop in water, especially well oxygenated and high quality water.

Habitat and Food Source(s)

The adults probably do not feed. The long mandibles of the males would make feeding very difficult. Hellgrammites are aggressive predators that attack other aquatic insects and other organisms in the water. They are found in flowing rivers and streams hidden among rocks. Immatures can be collected in riffle areas of streams where they hide among rocks (see Mayflies).

Pest Status

Adults are noticed when attracted to lights at night. Immatures (hellgrammites) are collected and used as fish bait; immatures are fierce biters, but medically harmless.

Dragonfly

Kingdom: Animalia, **Phylum**: Arthropoda, **Class**: Insecta, **Order**: Odonata, **Suborder**: Epiprocta, **Infraorder**: **Anisoptera** Selys, 1854

Families

- Aeshnidae
- Austropetaliidae
- Chlorogomphidae
- Cordulegastridae
- Corduliidae
- Gomphidae
- Libellulidae
- Macromiidae

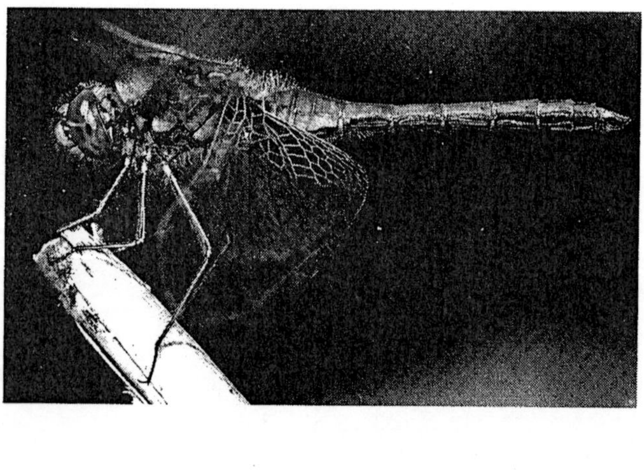

- Neopetaliidae
- Petaluridae
- Synthemistidae

A dragonfly is an insect belonging to the order Odonata, the suborder Epiprocta or, in the strict sense, the infraorder **Anisoptera**. It is characterized by large multifaceted eyes, two pairs of strong transparent wings, and an elongated body. Dragonflies are similar to damselflies, but the adults can be differentiated by the fact that the wings of most dragonflies are held away from, and perpendicular to, the body when at rest. Dragonflies possess six legs (like any other insect), but most of them cannot walk well. Dragonflies are some of the fastest insects in world.

Dragonflies are valuable predators that eat mosquitoes, and other small insects like flies, bees, ants, and very rarely butterflies. They are usually found around lakes, ponds, streams and wetlands because their larvac, known as "nymphs", are aquatic.

Female dragonflies lay eggs in or near water, often on floating or emergent plants. When laying eggs, some species will submerge themselves completely in order to lay their eggs on a good surface. The eggs then hatch into nymphs. Most of a dragonfly's life is spent in the nymph form, beneath the water's surface, using extendable jaws to catch other invertebrates (often mosquito larvae) or even vertebrates such as tadpoles and fish. They breathe through gills in their rectum, and can rapidly propel themselves by suddenly expelling water through the anus. Some nymphs even hunt on land, an aptitude which could easily have been more common in ancient times when terrestrial predators were clumsier.

The larval stage of large dragonflies may last as long as five years. In smaller species, this stage may last between two months and three years. When the larva is ready to metamorphose into an adult, it climbs up a reed or other emergent plant. Exposure to air causes the larva to begin breathing. The skin splits at a weak spot behind the head and the adult dragonfly crawls out of its old larval skin, pumps up its wings, and flies off to feed on midges and flies. In flight the adult dragonfly can propel itself in six directions; upward, downward, forward, back, and side to side. The adult stage of larger species of dragonfly can last as long as five or six months.

Life Cycle

Female dragonflies lay eggs in or near water, often on floating or emergent plants. When laying eggs, some species will submerge themselves completely in order to lay their eggs on a good surface. The eggs then hatch into nymphs. Most of a dragonfly's life is spent in the nymph form, beneath the water's surface, using extendable jaws to catch other invertebrates (often mosquito larvae) or even vertebrates such as tadpoles and fish. They breathe through gills in their rectum, and can rapidly propel themselves by suddenly expelling water through the anus. Some nymphs even hunt on land, an aptitude which could easily have been more common in ancient times when terrestrial predators were clumsier. The larval stage of large dragonflies may last as long as five years. In smaller species, this stage may last between two months and three years.

When the larva is ready to metamorphose into an adult, it climbs up a reed or other emergent plant. Exposure to air causes the larva to begin breathing. The skin splits at a weak spot behind the head and the adult dragonfly crawls out of its old larval skin, pumps up its wings, and flies off to feed on midges and flies. In flight the adult dragonfly can propel itself in six directions; upward, downward, forward, back, and side to side. The adult stage of larger species of dragonfly can last as long as five or six months

Habitat

Depending on the species of a dragonfly, the habitat also varies. Some species of the dragonfly habitat in marshy land, others prefer clean water or the still waters of the ponds. There are still other species, which have cool running waters of the streams or rivers. Dragonflies are choosy about their habitat. They would not reside in "polluted water. If you happen to see dragonflies around, it is a good sign that the ecosystem is in perfect and good condition. Remember, dragonflies are sensitive to pollution.

Importance

This dragonfly also is found in Asia, the Pacific islands, and the western hemisphere-from Alaska to Costa Rica. A common North American damselfly is the American ruby-spot, whose wings are white with a red spot near the base. Another striking species is the black-winged damselfly of the eastern United States. The male has black wings and the female dark gray and white wings. In both sexes, the body is metallic green. Dragonflies and damselflies are important in the food webs of freshwater streams and lakes. The immature forms live in freshwater as predators. They normally feed on other invertebrates, but some large species may occasionally eat fish or tadpoles. Adults consume large quantities of mosquitoes and other small flying insects. One of the most common dragonflies in the United States is the green darner, with a brilliant green head and thorax.

Earwig

Kingdom: Animalia, **Phylum**: Arthropoda, **Subphylum**: Hexapoda, **Class**: Insecta, **Order**: **Dermaptera** De Geer, 1773.

Earwigs make up the insect order Dermaptera, found throughout the Americas, Eurasia, Australia and New Zealand. It is one of the smaller insect orders, with only 1800 recorded species in 12 families. Typical earwigs have characteristic cerci, a pair of forceps-like pincers on their abdomen, and membranous wings folded underneath short forewings, hence the scientific name for the order, which translates literally as "skin wings".

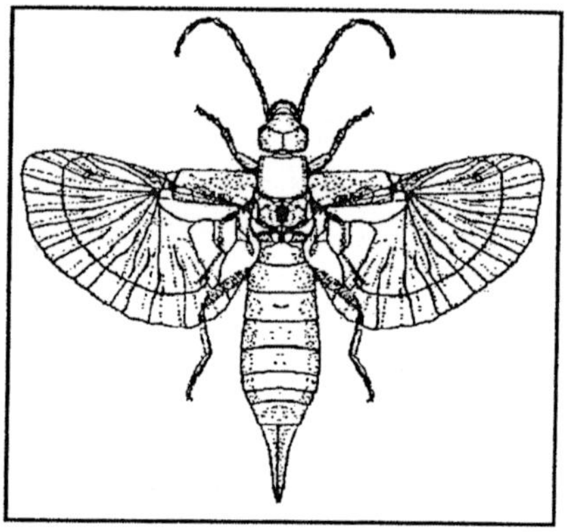

Some groups within the earwig order are tiny parasites on mammals and lack the typical pincers.

Earwigs rarely fly, even though they are capable of doing so. Earwigs are nocturnal; they often hide in small, moist crevices during the day, and are active at night, feeding on a wide variety of insects and plants. Damage to foliage, flowers, and various crops are commonly blamed on earwigs, especially the common earwig *Forficula auricularia.* Earwigs undergo an average of 5 molts over the course of a year, their average life expectancy, before they become adults.

Many earwig species display maternal care, which is uncommon among insects. Female earwigs are known to take care of their eggs, and even after they have hatched as nymphs will continue to watch over offspring until their second molt. As the nymphs molt, sexual dimorphism such as differences in pincer shapes begins to show. Earwig fossils have been found in different places. Some of those specimens are now included in the extinct suborder Archidermaptera dating back to the Late Triassic. Many orders of insect have been theorized to be closely related to earwigs by many authors, though Grylloblattaria is the most likely. Earwig are hemimetabolous, meaning they undergo incomplete metamorphosis, developing through a series of 4 to 6 molts.

Life Cycle

The developmental stages between molts are called instars. Earwigs live for about a year from hatching. They start mating in fall, and can be found together in the fall and winter. The male and female will live in a chamber in debris, crevices, or soil 2.5 mm deep. After mating, the sperm may remain in the female for months before the eggs are fertilized. From midwinter to early spring, the male will leave, or be driven out by the female. Afterward the female will begin to lay 20 to 80 pearly white eggs in 2 days. Some earwigs, those parasitic in the suborders Arixeniina and Hemimerina, are viviparous (give birth to live young); they would be fed by a sort of placenta. When first laid, the eggs are white or cream-coloured and oval-shaped, but right before hatching they become kidney-shaped and brown. Each egg is approximately 1 mm (0.04 in) tall and 0.8 mm (0.03 in) wide. Earwigs are among the few non-social insect species that show maternal care.

The mother will pay close attention to the needs of her eggs, such as warmth and protection, though studies have shown that the mother does not pay attention to the eggs as she collects them. The mother has been shown to pick up wax balls by accident, but they would eventually be rejected as they do not have the proper scent. The mother will also vigorously defend the eggs from predators, not eating unless an egg goes bad. Another distinct maternal care unique to earwigs is that the mother continuously cleans the eggs to protect them from fungi. Studies have found that the urge to clean the eggs persists for days after they are removed; when the eggs were replaced after hatching, the mother continued to clean them for up to 3 months.

The eggs hatch within seven days. The mother may assist the nymphs in hatching. When the nymphs hatch, they eat the egg casing and continue to live with the mother. The nymphs look similar to their parents, only smaller, and will nest under their mother and she will continue to protect them until their second molt in about July.

The nymphs feed on food regurgitated by the mother, and on their own molts. If the mother dies before the nymphs are ready to leave, the nymphs may eat her. After five to six instars, the nymphs will molt into adults. The male's forceps will become curved, while the females' remain straight. They will also develop their natural colour, which can be anything from a light brown (as in the Tawny earwig) to a dark black (as in the Ringlegged earwig). In species of winged earwigs, the wings will start to develop at this time. The forewings of an earwig are sclerotized to serve as protection for the membranous hindwings.

Habitat

They are especially common in the tropics and subtropics. Twenty-two species live in the United States and Canada. Most earwigs live in moist crevices (KREH-vuh-ses) of all kinds, including under bark, between leaves, and under stones. Some species live on the furry bodies of giant rats or bats as parasites, or animals that live on another organism or host and obtain food from it.

Food

Most earwigs are scavengers (SKAE-vihn-jers) and predators (PREH-duh-ters), feeding both on living and dead insects and plants. Some species eat mainly plants, while others eat mostly insects, such as chinch bugs, mole crickets, mites, scales, aphids, and caterpillars. Parasitic species scavenge bits of dead skin or fungi growing on the bodies of giant rats or feed on skin secretions of bats.

One species, the St. Helena earwig (*Labidura herculeana*), is listed as Endangered by the World Conservation Union (IUCN). It may even be extinct and no longer exist.

Flea

Kingdom: Animalia, **Phylum**: Arthropoda, **Class**: Insecta, **Subclass**: Pterygota, **Infraclass**: Neoptera, **Superorder**: Endopterygota, **Order**: **Siphonaptera** Latreille, 1825

Flea is the common name for insects of the order Siphonaptera which are wingless insects with mouthparts adapted for piercing skin and sucking blood. Fleas are external parasites, living by hematophagy off the blood of mammals (including bats and humans) and birds. Some flea species include:

- Cat flea (*Ctenocephalides felis*)
- Dog flea (*Ctenocephalides canis*)
- Human flea (*Pulex irritans*)
- Moorhen flea (*Dasypsyllus gallinulae*)
- Northern rat flea (*Nosopsyllus fasciatus*)
- Oriental rat flea (*Xenopsylla cheopis*)

Over 2,000 species have been described worldwide.

Life Cycle

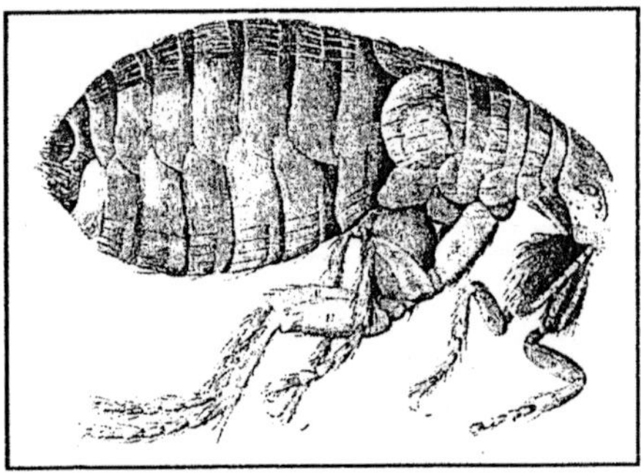

Fleas mate on their host animal and lay their eggs either onto the animal where they fall to the nest or directly in the nest. The small larvae hatch from the eggs and do not begin to feed on blood like that of their parents but consume the dead skin and other dirt and dust from the host animal. The larvae develop through 3 instars and when fully grown spin a silken cocoon and pupate in the nest of the host. The vibrations of a host often trigger the emergence of the adult flea from the pupal case, enabling it to immediately find a host and begin feeding. The complete life cycle may take from several weeks to many months depending on the species.

Feeding

Adult fleas feed on the blood of their host and although most fleas are adapted to one particular animal, many are not host specific and will move to other animals if their preferred host is unavailable. Fleas are very adaptable and are also able to withstand unfavourable conditions and can live for many days without food.

Habitat

Fleas are always found close by their hosts, either in direct contact such as among the feathers or hair or within their nests. Fleas are found in almost all habitats in Australia where there is a ready host and many native species of flea are closely associated with native marsupials and rodents. The main introduced species are also associated with animals that have been introduced to the country such as the cat flea (*Ctenocephalides felis*) the dog flea (*Ctenocephalides canis*) and the rat flea (*Xenopsylla cheopis*).

Fly

Kingdom: Animalia, **Phylum**: Arthropoda, **Class**: Insecta, **Superorder**: Panorpida, (**unranked**): Antliophora, **Order**: **Diptera** Linnaeus, 1758.

True flies are insects of the order Diptera (di = two, and ptera = wings). They possess a pair of wings on the mesothorax and a pair of halteres, derived from the hind wings, on the metathorax. The presence of a single pair of wings distinguishes true flies from other insects with "fly" in their name, such as mayflies, dragonflies, damselflies, stoneflies, whiteflies, fireflies, alderflies, dobsonflies, snakeflies, sawflies, caddisflies, butterflies or scorpionflies.

Some true flies have become secondarily wingless, especially in the superfamily Hippoboscoidea, or among those that are inquilines in social insect colonies. The common names of true flies are written as two words, e.g., crane fly, robber fly, bee fly, moth fly, fruit fly.

The common names of non-dipteran insects that have "fly" in their name are written as one word, e.g., butterfly, stonefly, dragonfly, scorpionfly, sawfly, caddisfly, whitefly. Diptera is a large order, containing an estimated 240,000 species of mosquitoes, gnats, midges and others, although under half of these (about 120,000 species) have been described.

It is one of the major insect orders both in terms of ecological and human (medical and economic) importance. The Diptera, in particular the mosquitoes (Culicidae), are of great importance as disease transmitters, acting as vectors for malaria, dengue, West Nile virus, yellow fever, encephalitis and other infectious diseases.

Life Cycle

The genitalia of female flies are rotated to a varying degree from the position found in other insects. In some flies this is a temporary rotation during mating, but in others it is a permanent torsion of the organs that occurs during the pupal stage. This torsion may lead to the anus being located below the genitals, or, in the case of 360° torsion, to the sperm duct being wrapped around the gut, despite the external organs being in their usual position. When flies mate, the male initially flies on top of the female, facing in the same direction, but then turns round to face in the opposite direction. This forces the male to lie on its back in order for its genitalia to remain engaged with those of the female, or the torsion of the male genitals allows the male to mate while remaining upright. This leads to flies having more reproduction abilities than most insects and at a much quicker rate.

Flies come in great populations due to their ability to mate effectively and in a short period of time during the mating season. The female lays her eggs as close to the food source as possible, and development is rapid, allowing the larva to consume as much food as possible in a short period of time before transforming into the adult. The eggs hatch immediately after being laid, or the flies are ovoviviparous, with the larva hatching inside the mother.

Larval flies, or maggots, have no true legs, and little demarcation between the thorax and abdomen; in the more derived species, the head is not clearly distinguishable from the rest of the body. Maggots are limbless, or else have small prolegs. The eyes and antennae are reduced or absent, and the abdomen also lacks appendages such as cerci. This lack of features is an adaptation to a food-rich environment, such as within rotting organic matter, or as an endoparasite.

The pupae take various forms, and in some cases develop inside a silk cocoon. After emerging from the pupa, the adult fly rarely lives more than a few days, and serves mainly to reproduce and to disperse in search of new food sources.

Habitat

Flies live in garbage and wherever animal feces are available. Dead animals attract flies within hours after death. Most flies are diurnal.

Food

Larvae feed on decaying meat and feces. Adult flies feed on sugary food of any kind, including nectar and rotting fruit.

BEE FLY

Kingdom: Animalia, **Phylum**: Arthropoda, **Class**: Insecta, **Order**: Diptera, **Suborder**: Brachycera, **Infraorder**: Asilomorpha, **Superfamily**: Asiloidea, Family: **Bombyliidae** Latreille , 1802.

Bombyliidae is a large family of flies with hundreds of genera, although their life cycles are not well known. Adults generally feed on nectar and pollen, thus are pollinators of flowers. They superficially resemble bees, thus are commonly called bee flies, and this may offer the adults some protection from predators.

The larval stages are predators or parasitoids of other insect eggs and larvae. The adult females usually deposit eggs in the vicinity of possible hosts, quite often in the burrows of beetles or wasps/solitary bees. Where most often in the insect world parasitoids are highly specific in the host species that they will infect, some bombyliids are opportunistic and will use a variety of hosts.

While bombyliids have a great variety of species, rarely are individuals of any one species abundant, and this is perhaps one of the poorest known families of insects relative to its size. There are at least 4,500 described species, and probably thousands as of yet undescribed.

Species

- Major bee-fly (Bombylius major)
- Williston's bee-fly (Poecilanthrax willistoni)
- Alkali bee fly (Heterostylum robustum).

Life Cycle

Larvae are parasitoids of the larvae of solitary bees such as *Andrena* species (Andrenidae).

Habitat

Woodlands and wood edges.

Food

Adults take nectar from flowers of herbaceous plants.

BLACK FLY

Kingdom: Animalia, **Phylum**: Arthropoda, **Class**: Insecta, **Order**: Diptera, **Suborder**: Nematocera, **Infraorder**: Culicomorpha, **Superfamily**: Chironomoidea, **Family**: **Simuliidae** Newman, 1834.

A black fly (sometimes called a buffalo gnat, turkey gnat, or white socks) is any member of the family Simuliidae of the

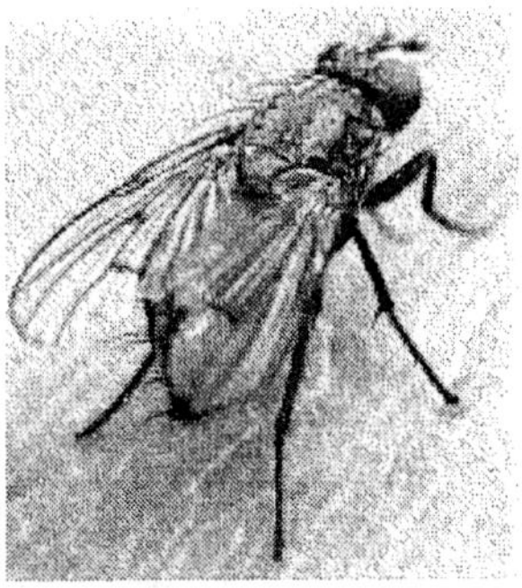

Culicomorpha infraorder. They are related to the Ceratopogonidae, Chironomidae, and Thaumaleidae. There are over 1,800 known species of black flies (of which 11 are extinct). Most species belong to the immense genus *Simulium*. Most black flies gain nourishment by feeding on the blood of other animals, although the males feed mainly on nectar. They are usually small, black or gray, with short legs, and antennae. They are a common nuisance for humans, and many U.S. states have programs to suppress the black fly population. They spread several diseases, including river blindness in Africa (*Simulium damnosum* and *S. neavei*) and the Americas (*Simulium callidum* and *S. metallicum* in Central America, *S. ochraceum* in Central and South America).

Life Cycle

Females deposit 150-500 creamy white eggs that darken until they are almost black just prior to hatching. The eggs usually are deposited singly or in masses in moving water or on some convenient object near the water's edge. The water may be a slow-moving stream, but most species prefer rapidly flowing water. In rare cases, breeding may occur in marshy areas. The time required for hatching varies with the species and may be 3-5 days or as long as 30 days at low temperatures. Larvae most often are found just beneath the surface of rapidly flowing water. They do not come to the surface to breath like mosquito larvae because the gill filaments of larvae extract oxygen directly from the water. Duration of larval development, like the egg stage, varies with species and temperature and ranges from 10-14 days to 7-10 weeks. Some species overwinter as larvae, although in northern areas, winter is usually passed in the egg stage. The last larval instar spins a reddish-brown, basket-like cocoon in which pupation occurs.

These cocoons are attached to slightly submerged objects such as rocks, logs, roots and other debris. Pupae also possess respiratory filaments with which they remove dissolved oxygen from the water. The pupal period varies from 4-5 days to as long as 3-5 weeks, depending on water temperature and species. Adults emerge from the pupal case, rise to the water surface, unfold their wings and fly away. Mating usually occurs shortly after the initial flight. The complete life cycle, from egg to adult, varies from 6-15 weeks and the number of generations per year ranges from one to six, depending on species and climatic conditions.

Habitat and Food Source(s)

Larvae are filter feeders and adult females have mouthparts modified for biting and blood feeding. Larvae of most species feed on small animals, such as protozoa and crustaceans, or plants such as algae, by straining particulate matter from water flowing by them with the aid of mouth brushes; some species feed in the silt or on submerged surfaces. Adults feed on the nectar of flowers; only the female requires a blood meal for ovarian development. Although a given species may prefer a particular animal host, most will readily feed upon other host species as well. Black flies attack man and a wide variety of domestic and wild animals and birds; others feed only on cold-blooded animals. Adults can be collected using a sweep net swung around ones head in areas and at times of the year when adults are active. Larvae and pupae can be found on rocks in the shallow water of waterfalls.

BLOW FLY

Kingdom: Animalia, **Phylum**: Arthropoda, **Class**: Insecta, **Order**: Diptera, **Suborder**: Brachycera, **Infraorder**: Muscomorpha, **Subsection**: Calyptratae, **Superfamily**: Oestroidea, Family: **Calliphoridae.**

Calliphoridae (commonly known as blow-flies, carrion flies, bluebottles, greenbottles, or cluster flies) are insects in the Order

Diptera, family Calliphoridae. The family is known to be non-monophyletic, but much remains disputed regarding proper treatment of the constituent units, some of which are occasionally accorded family status (e.g., Bengaliidae, Helicoboscidae, Polleniidae, Rhiniidae).

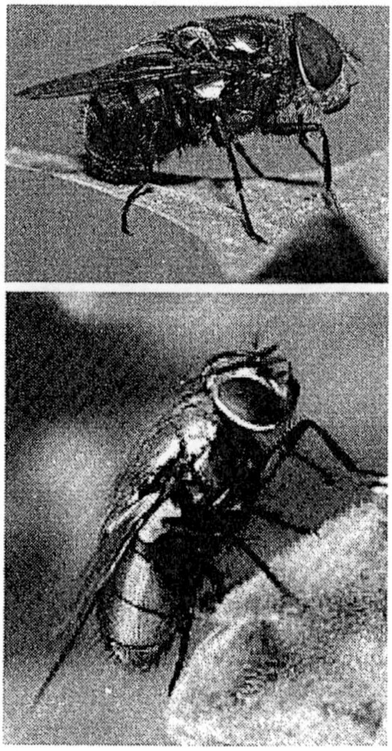

The name blow-fly comes from an older English term for meat that had eggs laid on it, which was said to be fly blown. The first known association of the term "blow" with flies appears in the plays of William Shakespeare: *Love's Labour's Lost*, *The Tempest*, and *Antony and Cleopatra*.

Life Cycle

Adults live between 9 - 21 days. Each species of blow fly is different, but blow flies can lay about 500 to 2200 eggs in their lifetime. Eggs are laid in batches of 100 to 200 in rotting meat and fish. Eggs are approximately 1.5 x 0.4 mm and look like tiny rice when laid. Besides the typical egg laying locations, garbage, manure, and rotting vegetable matter are also used.

Habitat

Blow flies are not usually a significant structural pest, but can be annoying due to their buzzing noise. Adult blow flies are occasional pollinators, being attracted to flowers with a strong odoue resembling rotting meat, such as the American pawpaw or Dead Horse Arum. There is little doubt that these flies utilize nectar as a source of carbohydrates to fuel flight, but just how and when this happens is unknown. Larvae of most blow fly species are scavengers of carrion and dung and most likely constitute the majority of the maggots found in such material, although it is not uncommon for them to be found in close associate with other dipterous larvae from other families.

BOT FLY

Kingdom: Animalia, **Phylum**: Arthropoda, **Class**: Insecta, **Order**: Diptera, **Section**: Schizophora, **Subsection**: Calyptratae, **Superfamily**: Oestroidea, **Family**: **Oestridae**

A botfly is a fly in the family **Oestridae**. It is one of several families of hairy flies whose larvae live as parasites within the bodies of mammals. There are approximately 150 known species worldwide. *Dermatobia hominis*, the human botfly, is the only species of botfly known to use humans as the host to its larvae.

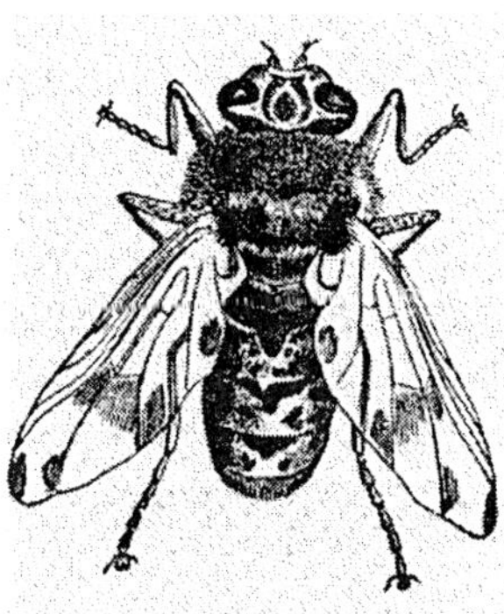

The sheep bot fly lays eggs in the nostrils of sheep. When hatched, the larvae can block the respiratory tract, thereby causing the animal's death. The human bot fly affects humans and members of the deer family in the Tropics. The female lays her eggs on mosquitoes and other biting insects, which carry them to the actual host. When the eggs come into contact with the warm host, the larvae hatch, burrow beneath the skin, and lodge in the muscles.

Bot flies are believed to be the fastest-flying of all insects and attain speeds of 64 to 80 km/h (40 to 50 mph). Although harmless in the adult stage, the larvae, called bots, are parasites that live in the body-cavity tissues of mammals, usually causing severe pain and sometimes death. The adult horse bot fly resembles a bee. It lays its eggs on the shoulders or flanks of the animal; the eggs are licked off by the horse and enter the digestive tract, where the larvae quickly hatch and attach themselves to the walls of the stomach and intestines.

Life Cycle

The adult female botfly catches a mosquito or muscoid fly. She does not kill the unlucky insect, but lays eggs on its body, according to the Centers for Disease Control and Prevention. The virtually microscopic eggs appear creamy in colour and oval in shape. The female botfly normally sticks her eggs on the sides of the blood-sucking insect's body. The eggs hatch into small larvae on the insect. The larvae remain on the host insect until it lands on a human or other mammal to take a blood meal.

Once the mosquito or muscoid fly begins to feed, the larvae leave the insect and quickly penetrate the skin of the human or other mammal through the bite hole left by the mosquito or muscoid fly. Rarely, the larvae will utilize a hair follicle to gain entrance into the skin. They will feed within the subdermal cavity they create just below the skin's surface. The larvae breath air using two barely noticeable posterior spiracles that stick out of the skin's surface.

The larvae will molt seven days after entering the host's body. They will continue to feed on the host's tissue for another 18 days before undergoing one last molt to reach their final size. The larvae will remain in the host's body for an additional 30 days before burrowing out of the host's skin and falling to the ground to pupate (transform into a pupa) within the soil.

The pupate stage of the botfly takes approximately two to three weeks. Once the pupate stage passes, the botfly emerges from the soil as an adult. It will quickly seek out another botfly to mate with before it perishes.

Habitat and Food Source(s)

Maggots have mouth hooks that tease tissue apart in the digestive tracts; adults do not have functional mouthparts. Horses, mules, donkeys are primary hosts. Adult female flies, attempting to lay eggs on host animals, cause horses to flee and resist fly "attacks" (hovering, buzzing and striking), occasionally resulting in injury. Larvae live in the digestive tract, injuring the tongue, lips, stomach lining and intestine. They apparently feed on the inflammatory products produced by the host in response to their presence. Infestations cause mechanical injuries and an infected ulcerous condition that progressively starves the host animal.

Control

Bots can be controlled with several types of dewormers, including dichlorvos, ivermectin and trichlorfon. In cattle, the lesions caused by these flies can become infected by *Mannheimia granulomatis*, a bacterium that causes lechiguana, characterized by rapid growing, hard lumps beneath the skin of the animal. Without antibiotics an affected animal will die within 3–11 months.

Certain type of botfly occasionally use humans as the host to its larvae. The larva, because of their spines, can pose an extremely painful sub-epidermal condition. Removal processes include placing raw meat on to the area, which in theory will coax the larva out. Another option is to use the tree sap of the *matatorsalo*, found in Costa Rica, which will kill the larva, yet leave its body in the skin. Additionally, one can attempt to seal the breathing hole of the larva with nail polish or vaseline and then, after a day, squeeze out the suffocated, dead larva. Use of adhesive tape can work, but carries additional risk of infection because portions of the larva's breathing tube can be broken off by the tape and made difficult to remove.

CHEESE FLY

Kingdom: Animalia, **Phylum**: Arthropoda, **Class**: Insecta, **Order**: Diptera, **Section**: Schizophora, **Subsection**: Acalyptratae, **Superfamily**: Tephritoidea, **Family**: **Piophilidae.**

Cheese flies are members of the family Piophilidae of flies. Most are scavengers in animal products and fungi. The best-

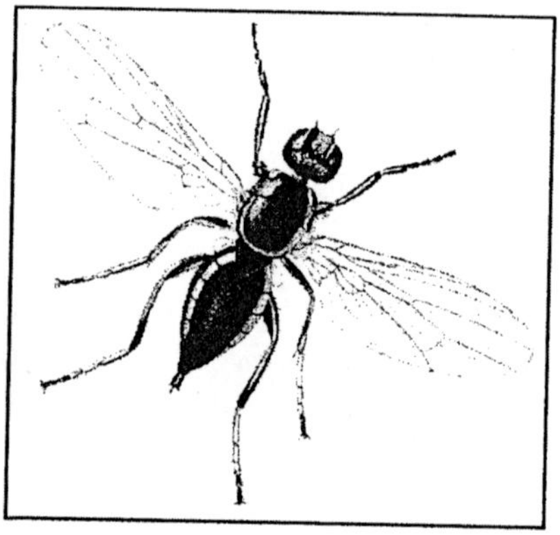

known member of the family is *Piophila casei*. It is a small fly, about four mm long, found worldwide. The fly's larva infests cured meats, smoked fish, cheeses, and decaying animals. The larva is about 8 mm long and is sometimes called the cheese skipper for its leaping ability - when disturbed, this tiny maggot can hop up to 15 cm into the air. Adults are also known as "bacon flies" and their larvae as "bacon skippers", "ham skippers", "cheese maggots", and "cheese hoppers". In Sardinia, Italy the larvae are intentionally introduced into pecorino cheese to produce casu marzu. If eaten (accidentally or otherwise), the larvae can pass through the digestive system alive (human stomach acids do not usually kill them) and live for some time in the intestines. This is referred to as an enteric myiasis.

Cheese fly larvae are a leading cause of myiasis in humans, and are the insect most frequently found in the human intestine. The larvae can cause serious intestinal lesions as they attempt to bore through the intestinal walls. Symptoms include nausea, vomiting, pain in the abdomen, and bloody diarrhea. Living and dead larvae may pass in the stool. Forensic entomology uses the presence of *Piophila casei* larvae to help estimate the date of death for human remains. They do not take up residence in a corpse until three to six months after death.

The larvae accomplish their jumps by bending over, grabbing onto the rears of their own bodies with their mouth hooks, tensing their muscles, and quickly releasing the grip. Spring action propels them into the air. A series of photos illustrating this remarkable behaviour in larvae of the tiny piophilid *Protopiophila litigata*, commonly known as the "antler fly" (it breeds exclusively on discarded antlers of cervids such as moose and deer), can be seen at. *Prochyliza xanthostoma* breeds on rotting carcasses and is highly sexually dimorphic, with males having greatly elongated heads, antennae and legs, relative to females. *P. xanthostoma* also has spectacular mating and fighting behaviours.

The adult fly's body is black, blue-black, or bronze, with some yellow on the head, antennae, and legs. The wings are faintly iridescent and lie flat upon the fly's abdomen when at rest. At four mm (1/6 inch) long, the fly is one-third to one-half as long as the common housefly. *Liopiopila* is a typical genus.

Life Cycle

Piophila casei undergo complete metamorphosis. Females mate almost immediately after adult emergence (Smith and Whitman 2000). Although feeding on protein sources increases fecundity of the insects, *P. casei* females are autogenous, able to produce eggs without a protein meal, in a laboratory setting. The complete life cycle of a cheese skipper in appropriate nourishment and temperature conditions can be as short as 12 days (1 day for egg development, 5 day larval maturation, 5 day pupal maturation, 1 day of adult feeding before reproduction). However, the typical life cycle is as follows (Mote 1914): Egg ~23 to 54 hours — Larva ~14 days — Pupa ~12 days — Adult ~3 to 7 days.

Distribution and Habitat

Cheese skippers have a worldwide distribution, including the United States, and are not limited to any specific geographic location.

Piophila casei usually feed on overripe (three or more months old) and moldy cheese, and slightly salted or putrid-smelling meats, such as ham, bacon, and beef. Larvae are typically found on high-protein substrates ranging from salted beef to smoked fish and animal carcasses. Signs of *P. casei* in foods include the presence of whitish-coloured eggshells as well as small grooves or creases found in the surface of cheeses made by first-instar larvae. Infested cheeses will usually have soft or sunken areas, and meats may have a shiny grease-like liquid drip from infested areas. Eggs, larvae, pupae, and adults are found near or on infested materials.

Importance

Medical

Cheese skippers are commonly cited as a cause of enteric (intestinal) myiasis in humans. Once a female oviposits (lays eggs) on meats, cheese, and other surfaces, the larvae hatch and penetrate deeply into the substrate. Unintentional human ingestion of cheese skipper larvae causes the maggots to pass through the human digestive system, often leading to serious intestinal lesions that result in diarrhea, pain, nausea, and other gastric symptoms. Cheese skippers are cited as one of the most problematic fly species associated with accidental myiasis. When myiasis occurs, *P. casei* larvae are usually found in the intestines, but larvae sometimes infest the chests and nasal passages of human patients. Because of their filth-feeding lifestyles, the adults are believed to be able to act as mechanical vectors of disease pathogens, similar to house flies.

Economic

Piophila casei are cited as pests of stored products ranging from salted meats to overripe cheeses. Damage to the stored products can incur both pest management costs and possible medical costs from accidental ingestion of the fly larvae.

Forensic

Forensic entomologists have used the presence of *P. casei* larvae as a tool to assist in the estimation of time of death for human remains. Though they can appear on remains less than two months old in geographic locations such as Florida, the flies sometimes do not appear on an exposed corpse until three to six months postmortem (after death), usually after the body has completed the "active decay" decomposition stage and is beginning to dry. Entomologists utilize knowledge of the current instar of collected larvae, coupled with measurements of weather and temperature conditions, to provide an estimation of the postmortem interval. Furthermore, unlike some other insects used in forensic investigation, the presence of drugs such as heroin do not significantly alter the development of *P. casei* larvae.

Management

Sanitation is considered the most important aspect of management of cheese skippers. Exclusion of *P. casei* from all levels of food production (processing, curing, and storage) ranges from proper sanitation techniques to proper wrapping/ protection of food products. Light infestations of *P. casei* larvae can be removed individually, but the judicious use of fumigation techniques if often required for more severe infestations.

Pachycrepoideus dubius Ashmead, a small pteromalid wasp, is the primary parasite of the cheese skipper. The wasp attacks the pupal stage of *P. casei,* as well as other cyclorrhaphan flies, including the Mediterranean fruit fly *Ceratitis capitata* (Wiedemann). *Pachycrepoideus dubius* is reported to provide effective control of the cheese skipper, especially during summer months. However, other research conducted at Ohio State University states that *P. dubius* does not act as an economically viable method for natural control of the cheese skipper. The small beetle *Necrobia rufipes* De Geer is also reported as a predator of cheese skipper larvae.

CRANE FLY

Kingdom: Animalia, **Phylum**: Arthropoda, **Class**: Insecta, **Order**: Diptera, **Suborder**: Nematocera, **Infraorder**: Tipulomorpha, **Superfamily**: Tipuloidea, **Family**: **Tipulidae.**

Crane flies are insects in the family Tipulidae. Adults are very slender, long-legged flies that may vary in length from 2–60 millimetres (0.079–2.4 in) (tropical species may exceed 100 millimetres or 3.9 in). In the United Kingdom, Ireland and Newfoundland they are commonly referred to as **daddy long-legs**, but this name can also refer to two unrelated arthropods: members of the arachnid order Opiliones (especially in the United States and Canada) and the cellar spider Pholcidae (especially in Australia). The larvae of the European Crane Fly are commonly known as Leatherjackets. These larvae can cause damage to lawns by feeding on the roots of grass plants. Numerous other common names have been applied to the crane fly, many of them more or less regional, including **mosquito hawk, mosquito eater** (or **skeeter eater**), **gallinipper**, and **gollywhopper**. At least 4256 species of crane flies have been described, most of them (75%) by the specialist Charles Paul Alexander. This makes the Tipulidae the largest family of Diptera.

Life Cycle

Adults emerge from soil between mid-August and mid-September. Females mate and lay eggs in the grass within 24 hours. Adults die without damaging grass. Eggs hatch into larvae, which feed in the fall and again in spring until about

mid-May. During June through August, larvae go just below the soil surface, stop eating, and pupate.

Habitat and Food Source(s)

Larvae have chewing mouthparts. Crane fly larvae feed primarily on decomposing organic matter. Adults do not feed. They commonly occur in moist environments such as woodlands, streams and flood plains although some species inhabit open fields, dry rangeland and even desert environments. In compost piles, they often occur on the soil surface below the pile of decaying vegetation. Some species have been reported to feed on roots of forage crops, turf grasses and seedling field crops. Usually their presence causes little concern because they are assisting in the process of decomposition. Larvae have not been reported to feed on vegetable transplants or garden plants in Texas. Adults have long slender legs which are easily broken and may be missing in some specimens.

Controls

Cultural and Physical

Maintaining proper irrigation, fertilization, and turf health helps the grass outgrow crane fly damage. If population levels near the action level are found, review and adjust cultural practices if needed. Aeration and dethatching in spring may help to reduce populations mechanically. Fall fertilization is very important. Optimum timing varies with fertilizer type: organic fertilizers need to be applied earlier and synthetic slow-release fertilizers later.

Biological Controls

Beneficial nematodes sold commercially have been shown to be effective in substantially reducing larval populations and may drop them far enough below the action threshold so that chemicals are not needed. Apply as directed in spring once soil temperatures have warmed to 55°F or greater. Turf must be kept irrigated to support the nematodes.

Chemical Controls

Dealing with crane flies without chemicals is environmentally preferred. EPA action in 2000 has severely limited chemical options for crane fly control. Chlorpyrifos is no longer allowed except on golf courses. Diazinon is illegal on golf courses and sod farms, and other outdoor uses are being phased out. Diazinon is highly toxic to fish and to birds that help control crane fly. Since many birds help to control crane fly, applying Diazinon is likely to be counterproductive in the long term. New chemicals are being evaluated for efficacy on crane fly.

FLESH FLY

Kingdom: Animalia, **Phylum**: Arthropoda, **Class**: Insecta, **Order**: Diptera, **Section**: Schizophora, **Superfamily**: Oestroidea, **Family**: Sarcophagidae.

Flies of the Diptera family Sarcophagidae are commonly known as flesh flies. Most flesh flies breed in carrion, dung, or decaying material, but a few species lay their eggs in the open wounds of mammals; hence their common name. Some flesh fly larvae are internal parasites of other insects. These larvae, commonly known as mag-gots, live for about 5-10 days, before descending into the soil and maturing into adulthood. At that stage, they live for 5-7 days.

Life Cycle

The life cycle of flesh-fly larvae has been well researched and is very predictable. Different species prefer bodies in different states of decomposition, and

the specific preferences and predictable life cycle timings allows forensic entomologists to understand the progress of decomposition and enables the calculation of the time of death by back extrapolation. This is done by determining the oldest larva of each species present, measuring the ambient temperature and from these values, calculating the earliest possible date and time for deposition of larvae. This yields an approximate time and date of death (d.o.d.) This evidence can be used in forensic entomology investigations and may assist in identification of a corpse by matching the calculated time of death with reports of missing persons. Such evidence has also been used to help identify murderers.

Habitat

Flesh flies are found in urban and rural communities but, fortunately, are relatively uncommon in houses or restaurants. They breed in excrement, decaying vegetable matter and animal flesh or meat.

Control

Flesh flies rarely require chemical control. The burial or destruction of carcasses, manure heaps and mounds of decaying organic matter eliminates many favorable breeding sites. Proper treatment of wounds also helps to prevent flesh fly harm to animals.

Exclusion and Sanitation, Insecticides, Traps, and Aerosols

1. Exclusion and Sanitation:
    - Drainage will often aid control, getting rid of extra moisture.
    - Garbage cans and dumpsters should have tight-fitting lids and be cleaned regularly.
    - Locate and eliminate all possible breeding sources. Flesh flies feed and breed on dead animals and garbage.Whenever possible, food and materials on which the flies can lay their

eggs must be removed, destroyed, as a breeding medium, or isolated from the egg-laying adult. Killing adult flies will reduce infestation, but elimination of breeding areas is necessary for good management

- Proper sanitation measures must be taken with dumpsters and
- rodent control with flesh flies.
- Exclude Flesh Flies from a structure with proper screening and maintenance of doors and windows.
- Openings of buildings should be tightly screened with screen.

FRUIT FLY

Kingdom: Animalia, **Phylum**: Arthropoda, **Class**: Insecta, **Order**: Diptera, **Family**: Drosophilidae, **Genus**: *Drosophila,* **Subgenus**: *Sophophora,* **Species group**: *melanogaster* group, **Species subgroup**: *melanogaster* subgroup, **Species complex**: *melanogaster* complex, **Species**: *Drosophila melanogaster.*

Fruit fly *(Drosophila melanogaster)* is a species of Diptera, or the order of flies, in the family Drosophilidae. The species is commonly known as the common fruit fly or vinegar fly. Starting from Charles W. Woodworth, this species is one of the most commonly used model organisms in biology, including studies in genetics, physiology, microbial pathogenesis and life history evolution because they are easy to take care of, breed quickly, and lay many eggs. Flies belonging to the family Tephritidae are also called fruit flies, which can lead to confusion, especially in Australia where the term fruit fly refers to the Tephritidae, economic pests in fruit production.

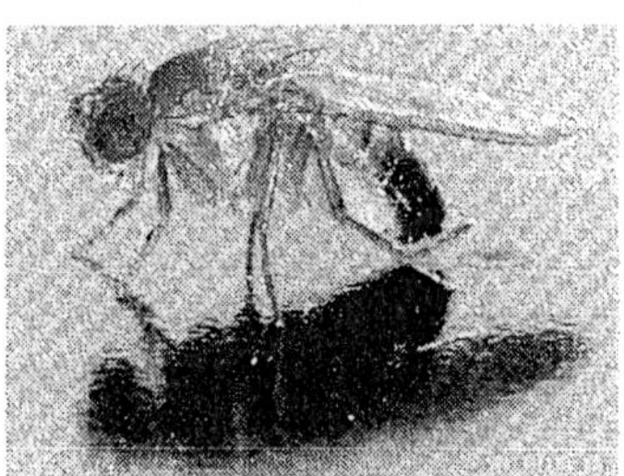

Life Cycle

There are four stages in the life cycle of the fruit fly: egg, larva, pupa, and adult. At a typical classroom temperature (21° C), *Drosophila melanogaster* will produce new adults in two weeks; eight days in the egg and larval stages, and six days in the pupal stage. The life span of a fruit fly may be several weeks. Twenty-four hours after the egg is laid, the larva hatches. The larva has two molting periods, during which the cuticle, mouth, hooks, and spiracles are shed. The larva is called an instar during the periods of growth before and after molting. Thus, the fruit fly has three instars. The puparium develops from the third instar which becomes hard in texture and dark in colour.

The puparium is where metamorphosis takes place. Just before the adult fly emerges from the puparium, the pupa darkens. Approximately 24 hours before the adult fly emerges, one can see the folded wings and the pigment of the eyes if one looks closely through the puparium. Upon completion of metamorphosis, the adult forces its way through the operculum (anterior end) of the puparium. Initially the fruit fly appears light in colour with a long abdomen and unexpanded wings. In just a few hours the fly gets darker in colour, rounder in the abdomen and extends its wings. Approximately 48 hours after emerging from the puparium, it is possible for females to start laying eggs. However, there are only 12 hours that a female can be considered virgin after emerging as an adult. After reaching this maturity, the flies are fertile for life. A female fruit fly can store sperm after a single insemination and use it for many reproductions; therefore it is important to use virgin females for genetic crosses.

Control

Mechanical Control

The two most common mechanical methods of control are wrapping developing fruit with a protective covering and the use of baited traps.

Cultural Control

The most effective cultural management technique is the destruction of all infested and unmarketable fruit, and the disposal of crop residues immediately after harvest.[*citation needed*]

Biological Control

Between 1947 and 1952, thirty-two species and varieties of natural enemies to fruit flies were introduced in Hawaii. These parasites lay their eggs in the eggs or maggots and emerge in the pupal stage. When the parasitoids F. arisanus or P. fletcheri were used, and attacked both melon fly eggs and larvae at the same time, suppression of development was as much as 56%.[5]

Irradiation

Sterilization of male melon flies through irradiation has proven effective in significantly reducing the number of eggs hatching.

Chemical Control

Toxicants in baits applied both to refugia of the fruit flies and sprays applied to crops have been used. Proteinaceous liquid attractants in insecticide sprays is an effective method of controlling melon fly populations. This bait insecticide is sprayed on broad leaf plants that serve as refugia for melon flies. These baits encourage the adults to feed on the spray residue.

GADFLY

"Gadfly", a fly that annoys horses and other livestock, usually a horse-fly or a botfly. A dipterous insect, belonging to the genus *tabanus* (Linn.), with three-jointed antennae and wide-spreading wings. The gadflies attack not only man, but cattle, horses, camels, and various ruminating animals. The most common species is the *T. atratus* (Fabr.), of a black colourof U.S.A., with a whitish bloom on the back, like that of a plum; the eyes are very large, almost meeting at the top of the head, and of a shining purplish or bronze black colour, with a jet-black band across the middle; it is about an inch long, with an expanse of wings of nearly two inches.

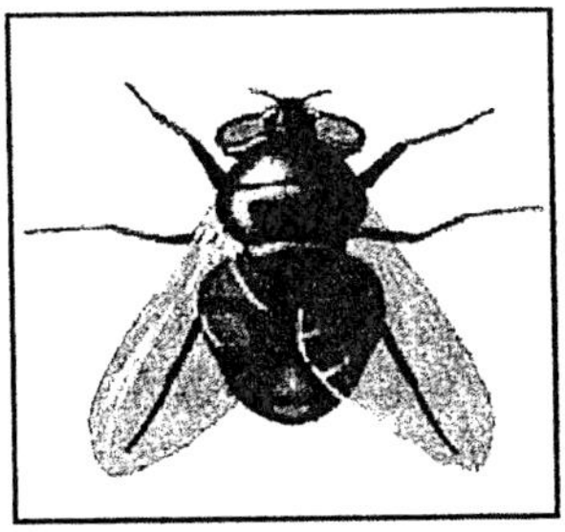

Orange-belted gadfly (*T. cinctus*, Fabr.) is smaller and less common, black, with the first three abdominal rings orange-coloured. The *T. lineola* (Fabr.) a smaller and very common species, which has a whitish line along the top of the hind body. There are about 40 European species. American species are *C. ferrugatus* (Fabr.), black, and *C. vittatus* (Wied.), striped black and yellow; they are found in woods and thickets in July and August. The genus *hoematopota* (Meig.) contains the troublesome gadfly called cleg in Scotland; this, the *H. pluvialis* (Meig.), is about the size of the common house, fly; the large eyes are greenish, with four undulating brown bands running through each of them; the body is gray, with brownish cross stripes; the wings gray, spotted with brown.

It attacks man, cattle, and especially horses, in sultry weather just before rain; the wounds are painful at the time, but are not followed by any lasting burning or itching. The name of gadfly is also given to several species of oestrus, especially to that which deposits its eggs about the knees and sides of the horse, and which, conveyed into the stomach, constitute the disease known as bots.

GALL MIDGE

Kingdom: Animalia, **Phylum**: Arthropoda, **Class**: Insecta, **Order**: Diptera, **Suborder**: Nematocera, **Infraorder**: Bibionomorpha, **Superfamily**: Sciaroidea, **Family**: **Cecidomyiidae.**

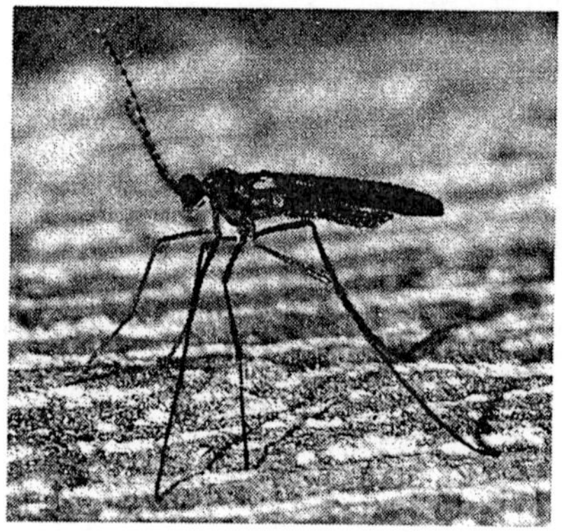

Cecidomyiidae (sometimes misspelled Cecidomyidae) is a family of flies (Order Diptera) known as gall midges or gall gnats. As the name implies, the larvae of most gall midges feed within plant tissue, creating abnormal plant growths called galls. These are very fragile small insects usually only 2-3 mm in length and many are less than 1 mm long. They are characterised by hairy wings, unusual in the Order Diptera, and have long antennae. Worldwide there are 3000+ species but since 1,100 are from well-studied North America this may be an underestimate.

Many are economically significant especially the important insect pest of wheat, the Hessian fly, the galls causing severe damage to the crop. Other important pests include the lentil flower midge (*Contarinia lentis*), the lucerne flower midge (*Contarinia medicaginis*) and the alfalfa sprout midge (*Dasineura ignorata*) on the Leguminosae; the Swede midge (*Contarinia nasturtii*) and the brassica pod midge (*Dasineura brassicae*) on the Cruciferae; the pear midge (*Contarinia pyrivora*) the raspberry cane midge (*Resseliella theobaldi*) on fruit crops; and the rosette gall midge (*Rhopalomyia solidaginis*) on goldenrod stalks. A large number of species are natural enemies of other crop pests.

The larvae of these species are predaceous, and some are even reported as parasitoids. The most common prey are aphids and spider mites, followed by scale insects , then other small prey such as whiteflies and thrips and many eat the eggs of other insects or mites. Because the tiny larva are incapable of moving considerable distances, there usually has to be a substantial population of prey present before the adults will lay eggs and Cecidiomyiidae are most frequently be seen during pest outbreaks. One species *Aphidoletes aphidomyza* is an important component of biological control programs for greenhouse crops and is widely sold in the United States of America.

Cecidomyiidae are also known for the strange phenomenon of paedogenesis in which the larval stage reproduces without maturing first. Even stranger in some species the daughter larvae produced within a mother larva consume the mother and in others the reproduction occurs in the egg or pupa.

Life Cycle

Gall midges spend the winter in the soil as a prepupa, an inactive stage between the larval and pupal stages when the insect develops into an adult. When the soil warms in the spring, the larvae develop into pupas, inside of which the young midge transforms into an adult. Adult midges lay their eggs on tender young rhododendron leaves, and the larvae feed upon the new leaf tissue. After a week, the larvae are mature and fall to the ground, where they spin a cocoon and mature into the next generation of adults. There are typically 3 or 4 generations of gall midges per summer.

Control

Two main Control methods,

1. Granular insecticides are recommended for gall midge control. Since farmers use granules after observing damage symptoms it is difficult to obtain a good control with granules.
2. Cultivation of resistant varieties in endemic areas is the most economical method.

Non-Chemical Control

Few non-chemical control methods for gall midge in rhododendron exist. Traditionally, gall midge was managed by periodically disturbing the soil and exposing the pupae to harsher above-ground conditions. Keeping plants healthy allows them to resist the damage caused by gall midge, and pruning infected leaves will stop the spread of the pest throughout the plant. Rhododendrons thrive in partial shade, protected from direct afternoon sun and wind and in well-drained, slightly acidic soil.

GNAT

A gnat (pronounced /Ènæt/ "nat") is one of several species of tiny flying insects in the Dipterid suborder Nematocera, especially those in the families Mycetophilidae, Anisopodidae and Sciaridae. Several nonbiting insects, such as the midges, which resemble mosquitoes, are also sometimes known as gnats. In North America the name is often applied to the black fly, midge, fungus gnat, biting midge, fruit fly, and other small flies that hover about the eyes of humans and other animals. In Great Britain the name usually refers to mosquitoes or less commonly to crane flies (family Tipulidae). Male gnats often assemble together in large mating swarms, particularly at dusk, called a *ghost*. Gnat larvae are mostly free-living and some are aquatic. Many feed on plants, though some are carnivorous. Larval plant feeders (e.g. the Hessian fly larva) cause root, stem, or leaf galls to be formed by the host plant. Some species of fungus gnats (families Mycetophilidae and Sciaridae) are very common pests of mushrooms and roots of potted plants in homes and greenhouses. Some South American Pleurothallid orchids are pollinated by tiny gnats and have correspondingly small flowers.

Life Cycle

The adult fungus gnat lays tiny eggs, singly or in clusters, on the soil in houseplants. In about 4 days, the eggs hatch into tiny, silvery white larvae (resembling tiny earthworms), with shiny black heads. When mature the very slim, and transparent larvae can measure up to 6 mm in length. Within a 2-week period, the larvae begin to spin tiny silken cocoons in the soil. The pupal stage lasts only 1 week. Adult fungus gnats emerge and mate. The adult stage lasts only 1 week, but the female will lay around 100-150 eggs during that time. The entire life cycle lasts only 4 weeks — from egg to larvae to pupae and back to egg.

Habitat

Gnats hatch from eggs and then spend up to two weeks as larvae. The larvae feed on decaying plant matter, such as leaf mulch. The flying gnats people notice can be found in most areas. They feed on decaying plant material, which is usually found in moist areas.

Control

One of the easiest ways to control gnats is to allow the soil in houseplants plants to dry before re-watering. Larvae cannot survive in dry soil. Gnats are most commonly found in soil that has a high percentage of organic matter (especially peat moss). Instead of using peat moss in, try perlite or vermiculite.

Insecticides have proven effective against fungus gnats. Diazinon (2% dust) applied to the soil surface and worked in slightly will come in contact with the larvae and kill them. Malathion (50% EC), at the rate of 2 ml per litre of water and applied to the soil surface is also effective. Note that some plants may be damaged by these insecticide. Read and follow directions carefully. Aerosol insecticides containing any of a number of chemicals (pyrethrins, resmethrin, tetramethrin, and d-trans allethrin) can be sprayed into the air. These insecticides must come in contact with the insect to kill it. Spray the soil, not the foliage, which may be damaged by the insecticide. Since these insects do not feed on the foliage, it is not necessary to spray the leaves the leaves.

Organic insecticides such as insecticidal soap and diatomaceous earth have proven effective in controlling fungus gnats, although further testing is required. Some plants may be damaged by these insecticides, therefore read and follow labels carefully. Brightly coloured traps have proven effective in some instances. One suggestion is to paint a milk carton bright yellow, then coat it with a sticky oil. Place or hang the carton near windows or where fungus gnats congregate. When the gnats are attracted to the bright yellow colour, they stick to the oil and die.

HESSIAN FLY

Kingdom: Animalia, **Phylum**: Arthropoda, **Class**: Insecta, **Order**: Diptera, **Family**: Cecidomyiidae, **Subfamily**: Cecidomyiinae, **Supertribe**: Lasiopteridi, Tribe: Oligotrophini, **Genus**: *Mayetiola,* **Species**: *M. destructor*

The hessian fly or barley midge, *Mayetiola destructor,* is a species of fly that is a significant pest of cereal crops including wheat, barley and rye. Though a native of Asia it was transported into Europe and later into North America, supposedly in the straw bedding of Hessian troops during the American Revolution (1775-83). There are usually two generations a year but may be up to five. In the spring the dark-coloured female lays about 250 to 300 reddish eggs on plants, usually where the stems are covered by leaves; the larvae feed on the sap and weaken the plants so that they cannot bear grain. The hessian fly was described by Thomas Say in 1817. It is a very harmful insect. It mainly attacks the stem, although if it is really hungry it will eat any part of the plant it can find. In 1836 a severe infestation of Hessian flies resulted in a crop shortage aggravating the financial problems of farmers prior to the Panic of 1837.

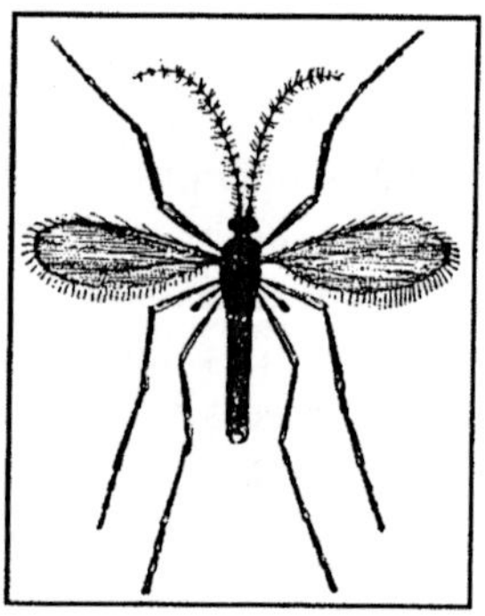

Life Cycle

Winter is spent in the pupal stage inside the puparium, just below the soil surface in the crown of the host plant. Adult flies emerge in spring and lay eggs on host plants. Larvae (maggots) hatch from eggs in 3 to 7 days, and develop through three stages (instars) over a period of 25 to 30 days before forming a puparium (by inflating the last larval skin). Summer is spent in field stubble. They are mainly active in the spring and fall, although there may be up to six generations per year.

Habitat

Larvae (maggots) secrete enzymes that cause stems to exude sap on which the larvae feed. Hessian flies attack small grains, including wheat, rye, barley and some wild grasses. Larvae (maggots) in leaf sheath tissue at the crown or joints along the stem, rupture plant cells and feed on sap. Injured plants are stunted and may die or fall over (lodge) before harvest. Adults are most easily reared from the pupal stages by placing infested crowns in a plastic bag.

Control

For a management program to be most effective, growers must implement a combination of all the techniques mentioned above in coordination with neighboring producers. For example, the efforts of a producer who rotates his wheat may be frustrated by a neighbor who plants wheat as a cover crop or who has a no-till double-crop soybean field adjacent to his farm. Careful selection of oats, rye, or triticale for cover cropping, avoiding planting wheat into or near old wheat stubble, the use of resistant varieties, and/or planting after the first frost will minimize Hessian fly damage.

HORSE FLY

Kingdom: Animalia, **Phylum**: Arthropoda, **Class**: Insecta, **Order**: Diptera, **Suborder**: Brachycera, **Infraorder**: Tabanomorpha, **Superfamily**: Tabanoidea, **Family**: Tabanidae

Insects in the order **Diptera**, family Tabanidae, are commonly called horse flies or deer flies. Often considered pests for the bites that many inflict, they are among the world's largest true flies. They are known to be extremely noisy during flight. They are also important pollinators of flowers, especially in South Africa. Tabanids occur worldwide, being absent only at extreme northern and southern latitudes. Flies of this type are among those known sometimes as gadflies, zimbs or clegs. In Australia, they are known as "March Flies". In some areas of Canada, they are also known as Bull Dog Flies. Elsewhere the term "horse fly" refers to the unrelated dipteran family Bibionidae. There are approximately 4,500 species of horse flies known worldwide, over 1000 of which are in the genus Tabanus. A three subfamilies are widely recognised:

- Chrysopsinae
- Pangoniinae
- Tabaninae
- the genus Zophina is of uncertain placement, though it has been classified among the Pangoniinae.

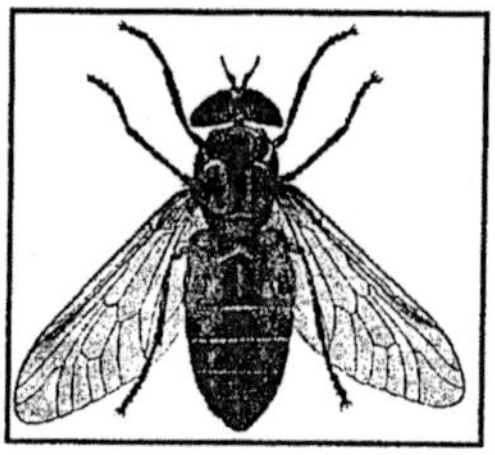

Two well-known types are the common horse flies, genus Tabanus Linnaeus, 1758 and the deer flies, genus Chrysops Meigen, 1802 are also known as banded horse flies because of their coloring. Both these genera give their names to subfamilies. The "Blue Tail Fly" in the eponymous song was probably a tabanid common to the southeastern United States.

Life Cycle and Habits

The life cycles of both flies are similar. Eggs are deposited in masses usually on vegetation or other objects over water near the larval habitat. After eggs hatch in 5 to 12 days, small larvae drop down and burrow into moist, wet soil found in marshes, stream banks, and bottoms of lakes and ponds. They may drop into rapidly flowing streams or burrow into dry soil. Larvae feed on organic debris, other insects, tiny crustaceans, snails, earthworms, and aquatic or semiaquatic organisms. Larvae overwinter in muddy soils, maturing in late spring. Pupation occurs in dry soil. The larval stage is about one year or up to two to three years for some species. The pupal period may range from 6 to 12 days depending on temperature and species. Adults are strong fliers, and appear in early summer with females feeding on blood while males feed on flower nectar, honeydew, plant juices, and other liquids. The life cycle may require from two months to two or three years, depending on the species and geographical region.

Habitat

Some species must obtain a blood meal before the development of each batch of eggs. Adults of most species are seen for only about one month, but often there is an overlapping succession of species during the season. Humans are seriously annoyed by deer flies around woodland areas with streams.

Control

No satisfactory methods have been developed for control of horse and deer flies. It is impractical in most regions to eliminate the breeding areas. Draining marshes and wet meadows where flies develop is of the greatest value, but should be done in such a way as to preserve the desirable wildlife of such areas, if possible. Fortunately, the season for deer flies is rather short, usually four to five weeks in June or July, and three to four weeks in August for horse flies. If the problem lasts six to eight weeks, there may be several species present or that species lasts longer than most. Humans are better able to protect themselves than wild or domestic animals by swatting flies away and by using repellents.

HOUSE FLY

Kingdom: Animalia, **Phylum**: Arthropoda, **Class**: Insecta, **Order**: Diptera, **Section**: Schizophora, **Family**: Muscidae, **Genus**: Musca, **Species**: *Musca domestica.*

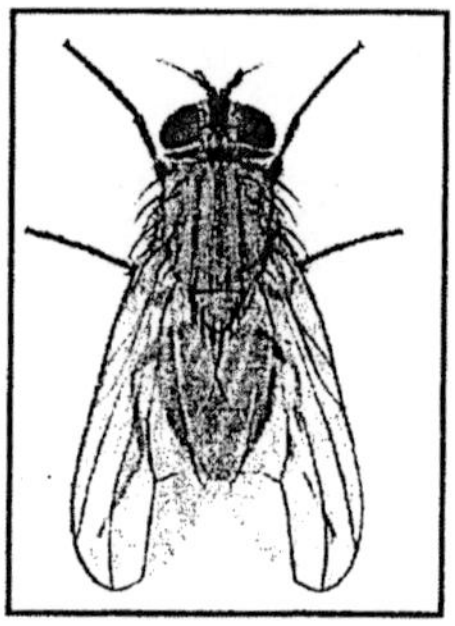

The housefly (also house fly, house-fly or common housefly), *Musca domestica,* is a fly of the Brachycera suborder. It is the most common of all domestic flies, accounting for about 90% of all flies in human habitations, and indeed one of the most widely distributed insects, found all over the world; it is considered a pest that can carry serious diseases.

Life Cycle

These are the four distinct stages in an average house fly's life:

- **Egg**: Depending on the size of a female house fly, she can lay up to 500 eggs in a three to four day period. Eggs are white in colour and are usually less than half and inch in size.
- **Larvae**: Larvae are commonly referred to as maggots. Maggots emerge from the eggs within eight to 20 hours of being laid. Larvae begin eating whatever they can find in the area they were laid. They prefer warm, moist environments to grow in.
- **Pupa**: After about four to 10 days, a maggot will move to higher, drier ground to move into the pupa stage of its life. This process take about three to six days and is where the maggot encases inself in a reddish-brown skin where the final stages of development take place.
- **Adult**: Once the adult house fly hatches from the pupal stage, it has an approximate life span of 15

to 30 days. Females are able to start producing eggs after two days of life and will continue to lay eggs for about a month. Female house flys are usually larger than the males.

Habitat

Near animal manure, garbage, or exposed food. Adult flies have sponging-sucking mouthparts, with which they inject mainly liquid food or food dissolved with regurgitated saliva. Larvae have mouthparts (mandibles) used to tease apart decomposing organic materials. Larvae feed with the ends of the bodies bearing the breathing pores on the surface and their narrow heads imbedded deep in the food source. Just before completing larval development, they leave their food source in search of a dryer place to pupate. This is the time many larval infestations become noticeable. Large numbers of house flies can develop in poultry houses and around barns and feed lots where animal excrement accumulates. House flies developing there can fly to nearby homes and become an nuisance. Around the home, house flies can develop in garbage and piles of fermenting lawn clippings.

Control

The house-fly nuisance can be abated most easily by the elimination of possible breeding-places. The great majority of the flies found in houses breed in piles of horse manure about near-by stables. Breeding in such places may be easily prevented by storing the manure, pending its removal, in a dark, fly-proof bin. This receptacle may be built as a lean-to attached to the stable with which it is connected by a small screen door. A larger door outside provides for the removal of the contents. The manure should be carted away at least once a week, and spread out on the land, where by drying it soon becomes unfit for breeding purposes. Whenever it is necessary to store such material in piles in the open, they should be located as far as possible from the nearest dwelling or milk-house.

Flies do not usually travel more than one-fourth mile from the place in which they breed. When only two or three horses are kept in a town, the manure can be handled in regular garbage-cans, in the same way as the kitchen refuse or ashes. It is rather difficult to treat manure piles with any substance to prevent breeding: chloride of lime, kerosene, and iron sulfate have been tried, but when used in economical quantities are not effective. Kitchen refuse and similar garbage should be kept in tight cans and removed at frequent intervals.

Flies should be rigidly excluded from all places where food is exposed to contamination, including kitchens, dining-rooms, stores, etc. Especial care should be taken to protect milk and milk utensils, since milk furnishes an excellent medium for the growth of typhoid bacteria and is a common source of infection.

Flies may be driven from rooms by leaving one door open and darkening all the rest. Then evaporate a spoonful of carbolic acid over a lamp, or burn some pyrethrum insect-powder. They may be caught on sticky sheets, or poisoned with a sweetened five per cent solution of commerical formaldehyde.

On isolated farms each owner has it in his power by proper measures in the disposal of manure to reduce the fly nuisance to a minimum. In towns the case is different; there cooperation is necessary.

MAGGOT

The legless, soft-bodied, wormlike larva of any of various flies of the order Diptera, often found in decaying matter. The living or rotting material furnishes heat for the hatching of the eggs and food for the newly hatched maggots. Certain blowfly maggots were once used in surgery to remove, by ingestion, both bacteria and the injured tissue in deep wounds.

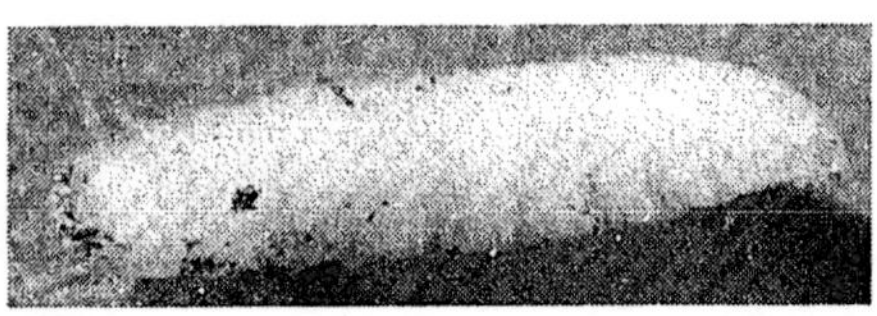

Life Cycle

Adult blow flies lay eggs on or near rotting vegetation, manure or animals. Eggs hatch very quickly, and maggots begin to consume the decaying food. Maggots pupate and emerge as adults within a few days, depending on the temperature. Total developmental time from egg to adult can happen as quickly as a week, given good conditions, and depending on the species. They develop and consume food quickly, so much so that when a maggot has matured, it moves away from the food source to digest and prepare for pupation. During this movement away from the food, a maggot can travel quite far, up to 100 yards. This is when people most often discover the maggots as they disperse through a wall and across the living room floor.

Habitat

They are usually the first critter to begin the process of decomposition and the recycling of trash, dead animals, and of course humans. A maggot mass can reduce the body weight of a corpse by 50% in a matter of a couple of weeks. There are many species of blow flies, and their lifecycles and habits are diverse. Each blow fly prefers certain seasons to be active and certain habitats. For example, in the cooler months like in fall and winter, species such as *Calliphora vomitoria* (nice name, huh?) are most commonly found on carrion, while *Phaenicia sericata* is commonly found out in the hot August days. Some flies even prefer shade over sun. Since blow flies are very specific in development and habitats, and they are the first insects to a corpse, they have proven to be the best crime investigators. A whole field of study has evolved from blow fly biology, called forensic entomology. Experts in this field use the fly maggot's identity, habitat preference, and developmental rate to determine the moment that someone was murdered and ultimately solve crimes. So, the next time you run across a filthy maggot.

Control

Generally white and resembling a worm or caterpillar, most maggots have a tendency to "gross out" even the toughest of men. In most cases you will see hundreds if not thousands at one location and the way they move makes it appear as though 10 times as many are actually present. Maggots are something most people don't like to see in and around their home. If you have seen them in your garbage or out in the yard, you know the feeling. Maggots are almost always the young of some type of insect. Most commonly the young of some specie of fly, maggots could be young beetles, moths or many other local and common insects.

Virtually all insects hatch out young which will start its life as a type of worm-like creature. Fly larva will almost always be white. They might have a tan, brown or black head but most people just see white. This is due to the sheer numbers that most people will happen upon when they first find any in or around the home. Since many insects will start out in this form, there is no common size nor location where they may be found. Maggots are generally associated with either garbage or a dead animal. However, they can readily feed on almost anything organic. This list includes but is not limited to carpeting, wallpaper, pet food, bird seed, pets, couches, clothing, furniture, pet hair, people hair, live animals, plants, fruit, vegetables, cooked meat or food, compost piles and just about anywhere in the home or immediately adjacent to it. Though maggots serve to "recycle" most any type of garbage or other decaying matter, most people don't want them in and/or around the house.

MEDITERRANEAN FRUIT FLY

Kingdom: Animalia, **Phylum**: Arthropoda, **Class**: Insecta, **Order**: Diptera, **Family**: Tephritidae, **Genus**: Ceratitis, **Species**: *Ceratitis capitata*.

Mediterranean fruit fly (*Ceratitis capitata*), or medfly for short, is a species of fruit fly capable of causing extensive damage to a wide range of fruit crops. It is native to the Mediterranean area, but has spread invasively to many parts of the world, including Australasia and North and South America.

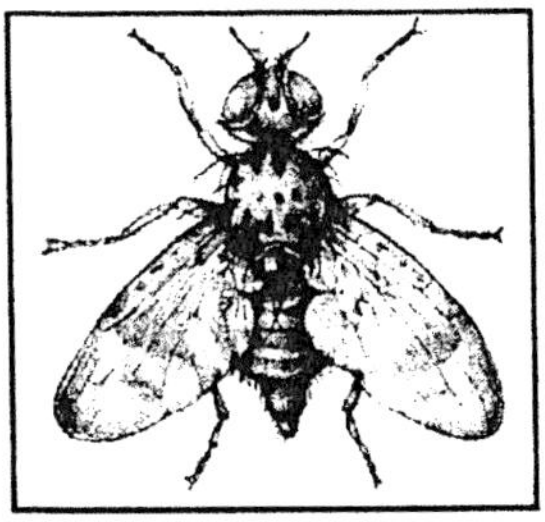

Life Cycle

Larva of the medflyAdult medflies lay their eggs under the skins of fruit, particularly where the skin is already broken. The eggs hatch within three days, and the larvae develop inside the fruit. The adults have a limited ability to disperse, but the global fruit trade can transport infected fruit over thousands of miles.

Habitat

The adult Mediterranean fruit flies are found wherever fruit trees grow. The larvae bore inside of fruit.

Control

It has been estimated that if control methods were not used, Medfly would infest 100 percent of susceptible fruit such as apricots, nectarines, peaches and mandarins and to a lesser extent, fruits such as apples and pears. All citrus is susceptible in warm winters. Only early maturing varieties of stone fruit or fruit fly tolerant varieties of fruit such as some lemon cultivars and avocados can be grown without insecticide applications. Two control methods are currently recommended: foliage baiting and cover spraying. Foliage baits are used to control the adult population; cover sprays control adult Medfly and provide some control of eggs and larvae.

MIDGE

Kingdom: Animalia, **Phylum**: Arthropoda, **Class**: Insecta, **Order**: Diptera, **Suborder**: Nematocera.

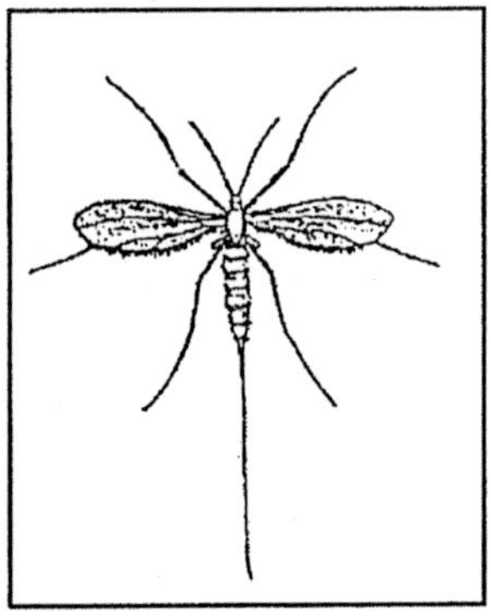

Midges comprise many kinds of very small two-winged flies found world-wide. The term does not encapsulate a well-defined taxonomic group, but includes animals in several families of Nematoceran Diptera. While some midges are vectors for disease, many others play useful roles as prey items for insectivores, such as frogs. The habits of midges vary greatly among the component families, which include:

- Blephariceridae, net-winged midges
- Cecidomyiidae, gall midges
- Ceratopogonidae, biting midges (also known as no-see-ums or punkies in North America)
- Chaoboridae, phantom midges
- Chironomidae, non-biting midges (also known as muffleheads in the Great Lakes region of North America)
- Deuterophlebiidae, mountain midges
- Dixidae, meniscus midges
- Scatopsidae, dung midges
- Thaumaleidae, solitary midges

Life Cycle

The midge begins life in the summer and it is the reproduction of the midge that is responsible for their biting behaviour. Once fertilised, the female's eggs cannot properly develop without a blood meal. For this reason she hunts down her prey - often us. One reason for the abundance of midges is that they produce two generations per season. The Highland

Midge is 'autogenus' - the female's first batch of eggs doesn't require a blood meal. Subsequent batches do.

A female midge will typically lay two batches of eggs. The first can contain up to 200 eggs, the second is usually smaller at up to 30. These are normally laid in moist soil, often by water. The larval midge emerges within a day or two and burrow into the ground. Under the right conditions just a few square yards of ground can contain hundreds of thousands of midge larvae.

The larvae go through four stages (instars), a process that takes about four weeks. Larvae from the first egg batch emerge later that same year, those from subsequent batches spend the winter underground. The fully developed midge larvae go through a short pupal stage of one or two days. In the case of second batch larvae that have spent the winter underground, this process is triggered by the warmer weather and usually takes place in May. Finally the adult midge emerges from the pupa, after which it has a lifespan of some 20 to 30 days in the wild. In laboratory conditions adult midges can live twice as long as this.

Habitat

Adults and larval dipterans live in nearly all habitats on land and in freshwater. Almost all adults are found on land and move about freely. However, louse flies spend most of their lives on their host animals. Dipteran larvae are found in fresh and brackish, or salty, water, wet soil and leaves, and other wet places. Some species live as external parasites on animals or bore through the tissues inside leaves. Larval shore flies live along the edges of hot springs and geysers where temperatures are more than 112°F (44.4°C). Others develop in pools of crude oil or inside the nests of ants, bees, and wasps. The pupae are usually found in the same habitats as the larvae. Aquatic species usually become pupae in the water, but species found along the water's edge sometimes prefer pupal sites away from water.

Control

Control of midges has been very ineffective. Nearly all attempts at control have yielded poor results at best because of the types of habitat that they breed in and their habits once they are adults. The problems with larval control are that, in general, the larvicides available are relatively ineffective against midges. This is probably because they are protected within their tubes in the soft mud and the pesticide is diluted by the large amounts of water in the breeding habitat. Likewise, adult control is also nearly as impossible. The adult midges tend to congregate on the lee-side of any obstruction (i.e., Homes, bushes, cars, etc.). Any type of area wide spray, even aerial spraying, does not reach the adults where they are swarming, and thus has no effect. An additional problem is that, since the adults are continuously emerging and replacing any that would be killed, any reduction would be short lived. Area type spraying has been tried in many areas around the country with virtually no success. All things considered, the most effective measures against midges are ones that the homeowners can do themselves.

MOSQUITO

Kingdom: Animalia, **Phylum**: Arthropoda, **Class**: Insecta, **Order**: Diptera, **Suborder**: Nematocera, **Infraorder**: Culicomorpha, **Superfamily**: Culicoidea, **Family**: Culicidae Meigen, 1830.

Mosquito (from the Spanish or Portuguese word for *little fly*) is a common insect in the family Culicidae (from the Latin *culex* meaning *midge* or *gnat*). Mosquitoes resemble crane flies (family Tipulidae) and chironomid flies (family Chironomidae), with which they are sometimes confused by the casual observer. Mosquitoes go through four stages in their life-cycle: egg, larva, pupa, and adult or imago. Adult females lay their eggs in standing water, which can be a salt-marsh, a lake, a puddle, a natural reservoir on a plant, or an artificial water container such as a plastic bucket.

The first three stages are aquatic and last 5–14 days, depending on the species and the ambient temperature; eggs hatch to become larvae, then pupae. The adult mosquito emerges from the pupa as it floats at the water surface. Adults live for 4–8 weeks. Mosquitoes have mouthparts that are adapted for piercing the skin of plants and animals.

While males typically feed on nectar and plant juices, the female needs to obtain nutrients from a "blood meal" before she can produce eggs. There are about 3,500 species of mosquitoes found throughout the world. In some species of mosquito, the females feed on humans, and are therefore vectors for a number of infectious diseases affecting millions of people per year.[8][9] Some scientists believe that eradicating mosquitos would not have serious consequences for any ecosystems.

Life Cycle

There are approximately 2700 species of mosquitoes world wide. The most common genera are Aedes, Culex, Culiseta, and Coquillettidia. Each genus may exhibit a slightly different lifecycle. However, all mosquitoes have the following stages: egg, larva, pupa, and adult.

Egg Stage

After the female has obtained her blood meal she will find a quiet place to rest and develop her eggs. It may take several days for this to happen. Once she is ready to lay her eggs she will seek out an appropriate place. The site selected and the way the eggs are laid largely depends on the genus. If she is in the genus Aedes she will most likely lay her eggs singly on the edge of a drying puddle or near the surface of water in a container. If she is in the genus Culex or Culiseta she will lay her eggs in a raft on the surface of the water. Each raft will contain several hundred eggs. She will repeat this cycle of obtaining blood meals and laying eggs until she dies. The duration of the egg stage is largely dependent on the species and on environmental conditions. The egg stage could last from one day to 9 months. Some mosquitoes overwinter as an egg. These eggs usually must experience a cold season and a specific day length to trigger hatching. For most species the eggs will hatch in one to seven days. When the larvae are ready to hatch they use a small temporary "tooth" on their head to break open the egg along a suture.

Larval Stage

Because the larva's skeleton is located on the outside (exoskeleton), similar to that of a crab, they must shed their exoskeleton in order to grow. All mosquito larvae shed their exoskeleton, or molt, four times; the stages between molts are called instars. Newly hatched larvae are called first instar larvaeand they are always very small and hard to see. Mosquito larvae typically float at the surface of the water. Here they can obtain food and breathe through their siphon. The siphon is located at the base of their abdomen and is similar to a snorkel. The larvae feed on bacteria and other organic matter in the water. Brushes that are located in front of their mouths collect the food.

Pupal Stage

After the larvae have completed their fourth instar stage they become pupae. This is the stage in which they undergo metamorphosis to become an adult mosquito. The process is similar to a caterpillar becoming a butterfly. The pupae are very active and look like commas. The mosquito will be a pupa for only a couple of days. While most insect pupae are inactive., mosquito pupae are unique because they are very active and can move quickly through the water. The pupae are transparent and the developing adult can be seen inside the pupal case.

Adult Stage

After one to three days the adult mosquito is ready to

emerge. The pupal skin splits along the top of the case. The adult mosquito slowly and carefully works its way out of the pupal case. After emerging it will float on the surface of the water and rest there until its body and wings harden. Once the body has hardened the mosquito will fly off to begin its new life. One of the first things newly emerged mosqitoes do is seek out nectar for a sugar meal to provide energy for flying and mating.

Generally male mosquitoes emerge a few days before female mosquitoes. This gives the males a chance to mature before the females emerge. The males then use their feathery antennae to hear the wings of the newly emerged females. Each mosquito species has a different sound to its wings so the males can find females of the same species. After they mate the female will look for a blood meal.

Only the adult female mosquito blood feeds. She needs the protein from the blood to develop her eggs. She obtains energy for herself from nectar. The male mosquito feeds only on nectar. All mosquito species do not blood feed on people. Some exhibit host preferences for birds or reptiles and amphibians while others do not blood feed at all

Most females die before they obtain their second blood meal but some may blood feed two or three times. Those females that obtain two or more blood meals transmit diseases since they have come in contact with several different hosts.

Habitat

Mosquitoes live around water. Various species breed and make their homes in the following areas:

Salt Marshes

Coastal salt marshes serve as breeding ground for several species of mosquitoes. Areas such as depressions and neglected ditches can breed millions of mosquitoes during the course of a summer. These mosquitoes can be effectively controlled through Open Marsh Water Management practices.

Freshwater Marshes

Specific species of mosquitoes are born in freshwater marshes. The larvae attach themselves to the stems and roots of the vegetation to obtain oxygen, and to acquire a certain amount of protection from predatory fish.

Temporary Woodland Pools

Shallow pools are common in wooded areas during the spring and wet summers. Various mosquito species will breed in this environment. They lay their eggs along the edges of the pools formed by rainwater.

Roadside Ditches

Roadside ditches are common habitats for many species of mosquitoes and ditches that hold water for extended periods of time become a breeding ground.

Freshwater Ponds

Some mosquito larvae is found primarily in small ponds where vegetated growth accommodates breeding.

Running Water

Streams with running water produce few, if any, mosquitoes. However, mosquitoes are sometimes found in isolated pockets of calm water away from the main flow of the stream.

Artificial Containers

Artificial containers like tires, buckets and gutters provide an ideal habitat for mosquitoes. Therefore, in order to keep the area around your home mosquito-free, all items collecting rainwater should be tipped over, removed or cleaned regularly.

Control

Mosquito control manages the population of mosquitoes to reduce their damage to human health, economies, and enjoyment. Mosquito control is a vital public-health practice throughout the world and especially in the tropics because mosquitoes spread many diseases, such as malaria. Mosquito-control operations are targeted against three different problems:

1. Nuisance mosquitoes bother people around homes or in parks and recreational areas;
2. Economically important mosquitoes reduce real estate values, adversely affect tourism and related business interests, or negatively impact livestock or poultry production;
3. Public health is the focus when mosquitoes are vectors, or transmitters, of infectious disease.

Disease organisms transmitted by mosquitoes include West Nile virus, Saint Louis encephalitis virus, Eastern equine encephalomyelitis virus, *Everglades virus*, Highlands J virus, La Crosse Encephalitis virus in the United States; dengue fever, yellow fever, Ilheus virus, and malaria in the American tropics; Rift Valley fever, Wuchereria bancrofti, Japanese encephalitis, chikungunya and malaria in Africa and Asia; and Murray Valley encephalitis in Australia. In the United States, states with sizable mosquito-control programs include California, Florida, New Jersey, Louisiana, Minnesota, Michigan, North Dakota, and Texas, among others.

ROBBER FLY

Kingdom: Animalia, **Phylum**: Arthropoda, **Class**: Insecta, **Order**: Diptera, **Suborder**: Brachycera, **Infraorder**: Asilomorpha, **Superfamily**: Asiloidea, **Family**: Asilidae

Insects in the Diptera family Asilidae are commonly called robber flies. The family Asilidae contains about 7,100 described species worldwide. All robber flies have stout, spiny legs, a dense moustache of bristles on the face (mystax), and 3 simple eyes (ocelli) in a characteristic depression between their two large compound eyes. The mystax helps protect the head and face when the fly encounters prey bent on defense.

The antennae are short, 3-segmented, sometimes with a bristle-like structure called an arista. The short, strong proboscis is used to stab and inject victims with saliva containing neurotoxic and proteolytic enzymes which paralyze and digest the insides; the fly then sucks the liquefied meal through the proboscis. Many species have long, tapering abdomens, sometimes with a sword-like ovipositor. Others are fat-bodied bumblebee mimics. Adult robber flies attack other flies, beetles, butterflies and moths, various bees, ants, dragon and damselflies, Ichneumon wasps, grasshoppers, and some spiders.

Life Cycle

Female Asilidae deposit whitish-coloured eggs on low-lying plants and grasses, or in crevices within soil, bark, or wood. Egg-laying habits depend on the species and their specific habitat; most species lay their eggs in masses, which are then covered with a chalky protective coating. Robber fly larvae live in the soil or in various other decaying organic materials that occur in their environment.

Larvae are also predatory, feeding on eggs, larvae, or other soft-bodied insects. Robber flies overwinter as larvae and pupate in the soil. Pupae migrate to the soil surface and emerge as adults, often leaving behind their pupal casing. Complete development ranges from one to three years, depending on species and environmental conditions. Theodor (1980) proposed that larval growth is accelerated in warmer regions and that many Asilidae species live no longer than one year.

Habitat and Food Source(s)

Adults have piercing-sucking mouthparts. Adult robber flies perch on stems of low plants or other objects and attack prey in the air. They feed on bees, beetles, dragonflies, other flies, grasshoppers, leafhoppers, wasps, and other insects. Larvae live in the soil, in wood and other habitats, feeding on organic matter, other arthropods such as white grubs, beetle pupae and grasshopper egg masses, and they may be carnivorous.

Control

Robber flies are important to nature and to people because they limit the populations of insect pests like flies, beetles, and grasshoppers. However, they cannot be considered strictly beneficial because robber flies also feed on insects like bees and dragonflies, and spiders which are also beneficial bugs.

Robber flies are not considered pests so insecticides are not necessary to control them. Robber flies can "bite" people, but do not do so unless they are trapped or confused. Don't try to capture a robber fly with your bare hands, in other words. Robber flies are interesting insects that can be fun to observe during the early summer in fields or meadows nearby a wooded area.

SAND FLY

Sand fly, any insect of the family Phlebotomidae (sometimes considered part of the family Psychodidae) of the order Diptera. The aquatic larvae live in the intertidal zone of coastal beaches, in mud, or in wet organic debris. Sand flies are of considerable medical importance: around the Mediterranean and in southern Asia, Phlebotomus transmits the pappataci fever virus; and in parts of South America, Africa, and Asia it carries the protozoan parasites causing kala azar, Oriental sore, espundia, and bartonellosis. The name sand fly is also used for certain species of the black fly and biting midge (qq.v) families.

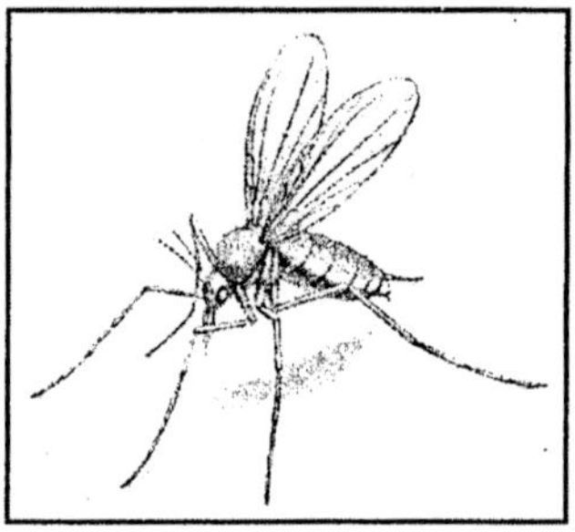

Life Cycle

The eggs are elongated oval-shaped, pale at first and darkening on exposure to air with a single black "eye spot". The larvae emerge through a J-shaped fissure and are legless and whitish with a dark head capsule. Those of the first instar can be distinguished by the presence of two caudal bristles, all subsequent instars bearing four. Fourth instar larvae also have a prominent sclerite on the dorsum of the penultimate segment. The pupae are golden brown and are affixed to the surface of the substrate in which they developed by the final larval exuvium. Shortly before emergence the wings and eyes turn black. Male sand flies emerge about 24 h before females, allowing their external genitalia time to rotate 180° to the correct position for mating before females have emerged. Although there have been no studies of sand fly development time in nature, the time from oviposition to adult emergence at ambient temperature is around 4-6 weeks. Some Palaearctic species diapause as larvae.

Habitat

Unlike in most biting Diptera, development of sand flies takes place in terrestrial rather than aquatic microhabitats. Although there have been relatively few successful attempts to identify breeding sites in nature, eggs are laid in soil rich in organic matter and the larvae pass through four instars before pupation and adult emergence. The difficulty of finding breeding sites is an important constraint on vector control measures available to leishmaniasis control programmes, application of larvicides not being a practical alternative.

Control

An initial evaluation of insecticide barrier spraying directed against sand fly vectors of cutaneous leishmaniasis was done in a nonclimax forested area with heavy undergrowth in Peten, Guatemala. A 100 m-wide swath of vegetation was sprayed once with a 1:3 mixture of cyfluthrin insecticide and a palm oil carrier using back-pack sprayers to simulate a central cantonment area in one site while another site remained as an untreated control. Prior to spraying and throughout 87 days post-treatment, sand fly populations were monitored at both sites with light traps set at ground and canopy levels at 50-m intervals radiating out from the centers of the cantonments, 150-m in the four cardinal directions.

A total of 2,876 female sand flies were captured, representing 16 species. Three species, *Brumptomyia galindoi, Lutzomyia panamensis,* and *Lu. ovallesi,* comprised 70% of the total collection.

The single insecticide barrier significantly reduced sand flies from reaching the cantonment area for more than 80 days, while sand fly populations outside the treated cantonment and in the untreated (control) cantonment remained high (52 sand flies in the treated cantonment versus 235 sand flies in the untreated cantonment).

SCREW-WORM

Kingdom: Animalia, **Phylum**: Arthropoda, **Class**: Insecta, **Order**: Diptera, **Family**: Calliphoridae, **Genus**: *Cochliomyia* Species: *Cochliomyia hominivorax.*

Cochliomyia hominivorax, the New World screw-worm fly, or screw-worm for short, is a species of parasitic fly that is well known for the way in which its larvae (maggots) eat the living tissue of warm-blooded animals. It is present in the New World tropics. There are five species of Cochliomyia but only one species of screw-worm fly in the genus; there is also a single Old World species in a different genus (Chrysomya bezziana). Infestation of a live vertebrate animal by a maggot is technically called myiasis. While the maggots of many fly species eat dead flesh, and may

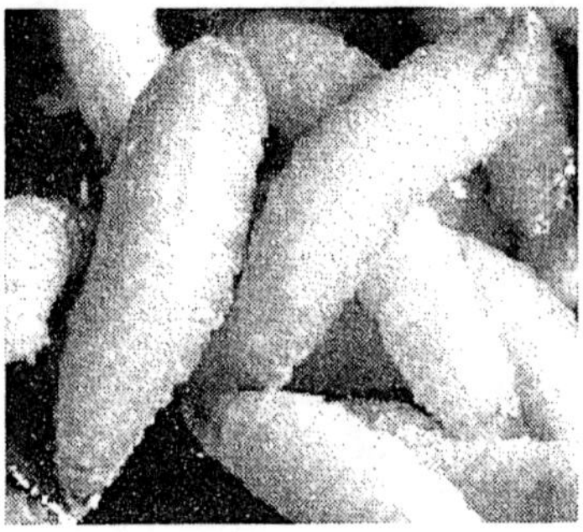

occasionally infest an old and putrid wound, screw-worm maggots are unusual because they attack healthy tissue. Screw-worms are a reportable species to the state veterinarian in the United States if discovered on livestock.

Life Cycle

Screw-worm females lay 250-500 eggs in the exposed flesh of warm-blooded animals, including humans, such as in wounds and the navels of newly-born animals. The larvae hatch and burrow into the surrounding tissue as they feed. If the wound is disturbed during this time the larvae burrow or "screw" deeper into the flesh, which is the source of the insect's name. The maggots are capable of causing severe tissue damage or even death to the host. Approximately three to seven days after hatching the larvae fall to the ground to pupate. The pupae reach the adult stage about seven days later. Female screw-worm flies mate four to five days after hatching. The entire life cycle is approximately twenty days. A female can lay up to 3,000 eggs and fly up to 200 km (125 miles) during her life.

Habitat

The habitat was a low deciduous forest comprised predominantly of trees *ca* . 3–8 m tall, with scattered, emergent trees above the canopy. The study site at Finca San Rafael was also situated in a low deciduous forest, although the trees were taller (ca . 4–10 m) and the vegetation was more dense.

Control

The United States officially eradicated the screw-worm in 1982 using the sterile insect technique. The screw-worm was eradicated in Guatemala and Belize in 1994, El Salvador in 1995, and Honduras in 1996. Campaigns against the flies continue in Mexico, Nicaragua, Costa Rica, Panama and Jamaica with financial assistance from the United States Department of Agriculture.

TSETSE FLY

Kingdom: Animalia, **Phylum**: Arthropoda, **Class**: Insecta, **Order**: Diptera, **Subsection**: Calyptratae, **Superfamily**: Hippoboscoidea, **Family**: Glossinidae Theobald, 1903, **Genus**: Glossina Wiedemann, 1830.

Tsetse sometimes spelled tzetze and also known as tik-tik flies, are large biting flies that inhabit much of mid-continental Africa between the Sahara and the Kalahari deserts. They live by feeding on the blood of vertebrate animals and are the primary biological vectors of trypanosomes, which cause human sleeping sickness and animal trypanosomiasis, also known as nagana. Tsetse include all the species in the genus Glossina, which are generally placed in their own family,

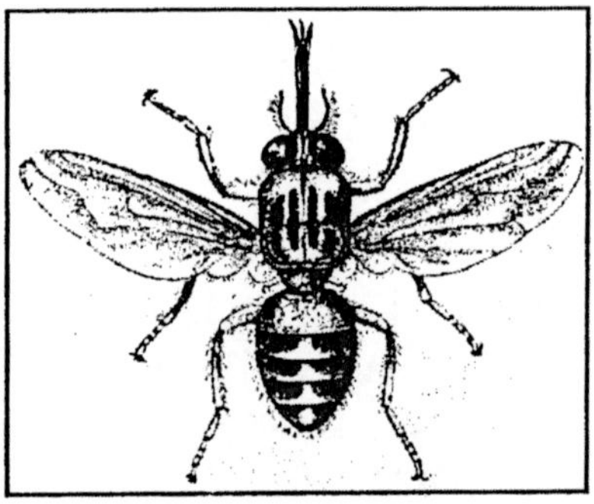

Glossinidae. Tsetse have been extensively studied because of their disease transmission.

These flies are multivoltine, typically producing about four generations yearly, and up to 31 generations total over their entire lifespan. Tsetse are crudely similar to other large flies, such as the housefly, but can be distinguished by various characteristics of their anatomy, two of which are easy to observe. Tsetse fold their wings completely when they are resting so that one wing rests directly on top of the other over their abdomen. Tsetse also have a long proboscis, which extends directly forward and is attached by a distinct bulb to the bottom of their head. Fossilized tsetse have been recovered from the Florissant Fossil Beds in Colorado showing that tsetse have existed in their modern morphological form for at least 34 million years. There are 23 species of tsetse flies. Diseases transmitted by tsetse flies kill 250,000–300,000 people per year.

Life Cycle

The female tsetse fly does not lay eggs but produces larvae, one at a time. The larva develops in the uterus over a period of 10 days and is then deposited fully grown on moist soil or sand in shaded places, usually under bushes, fallen logs, large stones and buttress roots. It buries itself immediately and turns into a pupa. The fly emerges 22–60 days later, depending on the temperature. Females mate only once in their life and, with optimum availability of food and breeding habitats, can produce a larva every 10 days.

Habitat

Tsetse flies are found in patches of dense vegetation along banks of rivers and lakes in hot, dry habitats. They also live in dense, wet, heavily forested equatorial rainforest.

Control

Tsetse control has been undertaken to reduce the incidence of the diseases the flies transmit. Two alternative strategies have been used in the attempts to reduce this African trypanosomiasis. One tactic is primarily medical or veterinary, targeting the disease directly using monitoring, prophylaxis, treatment, and surveillance to reduce the number of organisms that carry the disease.

The second strategy is generally entomological, and intends to disrupt the cycle of transmission by reducing the number of flies. The idea of tsetse control implies a change in the relationship between people and these insects. Prior to the twentieth century, people in Africa had largely adapted to the presence of tsetse. Human settlement patterns and agricultural practices had adapted to the presence of the fly. For example, in Ethiopia draft powered farming was restricted to the highland areas where the flies were absent, whereas lowland areas where tsetse are present were more sparsely populated by people living a nomadic, less agriculturally intensive lifestyle. Tsetse control is a response to changing conditions.

Tsetse control has been proposed as a way of reducing the incidence of the disease in the populations living in tsetse regions, of allowing the expansion of human settlement and agriculture into new areas, and of helping people previously relocated either in forced transfers or due to migration. Tsetse control efforts have been undertaken throughout the African continent, but long-term, sustainable control has rarely been achieved.

Tsetse control efforts invariably are tied to the complex problems of poverty, health, politics, and violence that have proved so disastrous for the African people. The reduction of fly numbers has generally been attempted with two different aims, either total eradication from the area, or control to just reduce the numbers. Eradication has often been imagined, has repeatedly been attempted, and is still proposed--but many reasons suggest that control is a safer, cheaper, more realistic, and sustainable approach. Eradication refers to the successful killing of every tsetse, either in a region or, under more grandiose proposals, from the entire African continent.

Local eradication efforts have repeatedly been undertaken and have achieved temporary success, only to fail in the long term because tsetse re-invaded (as in Zanzibar). All of the economic, ecological, political, and environmental justifications for eradication have been called into question. The economic justification for eradication offsets the immense costs of the eradication campaign against the medical and veterinary benefits that accrue in perpetuity.

Grasshopper

Kingdom: Animalia, **Phylum**: Arthropoda, **Subphylum**: Hexapoda, **Class**: Insecta, **Order**: Orthoptera, **Suborder**: Caelifera Ander, 1939.

The grasshopper is an insect of the suborder Caelifera in the order Orthoptera. To distinguish it from bush crickets or katydids, it is sometimes referred to as the short-horned grasshopper. Species that change colour and behaviour at high population densities are called locusts. Grasshoppers are vegetarian insects, which can jump, walk and fly. Grasshoppers have a typical 3 - part insect body, which includes the head, thorax and abdomen. They also have 2 antennae, 2 pairs of wings and 6 legs. Keep reading to learn about more about grasshoppers habitat and facts about grasshoppers. Grasshoppers can eat almost anything, from clothes, tomatoes, paper, and even toxic plants! Maybe that is the reason for their brilliant adaptability which in turn has resulted in their occupation of varied grasshoppers habitats.

Some grasshopper facts include:

- They can shed their skin four or five times

- There are 18,000 different types of grasshopper in the world
- All Grasshoppers are Herbivores
- They have five eyes
- They do not have ears
- They can leap 20 times the length of their own bodies
- When picked up they spit out a liquid which is commonly called tobacco juice
- They can survive almost anywhere

As mentioned, grasshoppers are Herbivores. So exactly what do grasshoppers eat?Grasshoppers aren't fussy with what they eat. They eat grass, vegetable plants and flowers. It is common for swarms of grasshoppers to eat farmer's crops which cause extreme damage for the farmer.

Life Cycle

A grasshopper's life cycle revolves around eggs, nymphs and adult stages. These insects belong to the order Othoptera. There are nearly 11,000 different species of grasshoppers found all over the world. They are predominantly green in color and feed on leaves and grasses of other plants. They have powerful hind legs and wings attached to their thorax. The fore wings are leathery and somewhat narrow. The hind wings on the other hand are large and membranous.

Females are bigger than males. The males attract the females by singing a melodious tune. After mating, the females carry the eggs in a special sac like structure. Reproduction mainly occurs in early summer. By mid summer, the female will lay her eggs on sand or littered leaves. The female uses a pair of valve like triangular structures on her abdomen for digging sand. Next, she will lay her eggs and allow nature to take its own course. The eggs are laid as egg pods. Each pod has nearly 30 to 100 rice shaped eggs.

These eggs hatch in autumn. Nymphs emerge out and look pretty much like smaller sized versions of adult grasshoppers.

They do not have wings. They feed on moist leaves scattered on the ground and molt several times. They increase in size and after a period of 20 to 30 days, they will mature into adults. The adults reach maturity and become sexually active in a fortnight. They survive for around 30 days. A grasshopper has a lifespan of one year.

Habitat

Grasshoppers are observed in all kinds of habitats. Some examples of grasshoppers habitat are grasslands, rain forests, pastures, fields, swamps and tropics. Most surprisingly they are also found in deserts. In case of adverse conditions, grasshoppers are known to migrate in large numbers. Areas with dense vegetation like scrubs, are home to some rare and interesting species of grasshoppers.

Common Grasshoppers Habitat

Grasshoppers habitats are generally concentrated in areas that have low growing plants, since they like to eat grass, leaves, and cereal crops. But, if there is depletion of these foods they start eating other favorable food for survival.

Rare Grasshoppers Habitat

Grasshoppers are rare in freshwater swamps and mangrove swamps because of the destruction they cause to the eggs that the insects lay in the soil, due to flooding.

Habitat Manipulation of Grasshoppers

Grasshoppers feed on crops like barley, corn, oats, etc., which is why they are regarded as pests by farmers. Habitat manipulation of grasshoppers is a control strategy, where the number of natural predators of grasshoppers are increased in the fields. This helps in depletion of grasshopper population, and thus protects the crops.

Control

In order to control grasshoppers you need to first see how big your problem is. If there are only a couple of grasshoppers, it is more than likely that you can leave them alone as they do not pose as a threat. However they can create havoc and so there are a number of grasshopper control strategies that you could use.

These strategies include:

- Planting seeds early
- Using Insecticides
- Baits
- Keeping Poultry

Most people tend to use insecticides to get rid of grasshoppers and other garden pests. However you have to make sure you read the label carefully before applying the insecticide over the infected area. Insecticides can cause a lot of damage and great care needs to be taken when using them. However they are highly effective though several treatments may need to be used. By planting seeds earlier than usual, crops and plants get the chance to grow more before grasshoppers make an appearance. Whilst this doesn't stop damage altogether, older plants are better protected against grasshopper attacks than younger plants. Baits are also effective and can be easily bought but they can be slightly expensive. They are best placed around when the grasshoppers are still young. They usually contain bran and some form of insecticide.

If you have a problem in your garden you may want to think of getting some form of poultry. Chickens for example will eat roaming grasshoppers and it will deter them from the area. Grasshoppers are real pests and they come in large numbers. They can be tricky but there are plenty of products to help you eliminate the problem.

Grylloblattid
(Rock Crawlers, Ice Crawlers)

Class: Insecta, **Phylum:** Arthropoda.

Grylloblattids (also known as ice crawlers or rock crawlers) are a poorly known group of insects named for their resemblance to both crickets and cockroaches. They are restricted to cold and extreme habitats such as glaciated mountains or ice caves. Grylloblattids are found only in Japan, Siberia, the northwestern United States, and western Canada. There are 26 species known worldwide; the 10 North American species are restricted to icy mountains in Montana, California, Oregon, Washington, and western Canada. As glaciers and ice fields recede due to the effects of climate change many grylloblattid habitats are threatened. Therefore, Grylloblattodea stands as the least known and the most threatened insect order.

The Grylloblattodea, or ice crawlers, are a small group of soft-bodied, apterous, terrestrial and termitelike insects confined to the Northern Hemisphere. They occur under rocks in forest leaf litter or above the treeline in the high mountains. In parts of the United States, Canada, and China they are found under rocks in soil at or below freezing temperature. In Japan and Korea, they are active in midsummer in the deep leaf litter of mixed conifer and deciduous forests, where daily ambient temperatures approach 30°C. In Japan and Korea some have been found in caves; in the western United States, they are known from subterranean lava tubes.

Grylloblattids are slender, depressed insects covered with fine hairs and having reduced eyes. Adults range from 2 to 3.5

cm in length. Ice crawlers appear to be primarily nocturnal. They are considered to be some of the most primitive of orthopteroid insects and have been thought to be related to cockroaches (Blattodea) and stick insects (Phasmida).

Classification and Features of the Order

The Grylloblattodea comprise a single family, Grylloblattidae, which includes only four genera. Grylloblatta occurs in North America and Canada, where at least 11 species are known. Grylloblattina is known from a few species from Siberia. Grylloblattella has species known from Korea, Japan, and Siberia. Galloisiana contains at least 10 species from Japan. The group, however, is undoubtedly larger, since new species are discovered every few years. Grylloblattids are cryptic, ground-dwelling insects that prefer wet habitats and cool temperatures.

They shun light and occur under stones or in dense leaf litter. Species of Grylloblatta that live under rocks under or near snowbanks emerge after dark and feed as scavengers or predators on dead or dying insects that have been blown onto the snow from lower elevations. During winter, the species probably occupy the airspace between the ground and overlying snowpack, where they remain active at temperatures of 0°C. Massive fat bodies build up prior to winter; during winter, the insects may feed on decaying plant material.

Korean species live under debris on the floor of caves at only 200 m altitude and apparently never venture forth from the cave habitat. The most widespread species in Japan, Galloisiana nipponensis, is found at elevations ranging from 300 to 3000 m, where the insects live under stones and in the leaf litter of thick, mixed coniferous and hardwood forests. These species consume both insects and plant material day and night. The resemblance of grylloblattids to Dermaptera has been shown to be superficial and is associated only with the fact that both have a projecting head. The head, however, is typical of orthopteroid insects. The antennae are elongate and thin. The mouthparts are structured like those of a predator.

Ocelli are absent, and the eye comprises fewer ommatidia in young instars than in adults. The legs are simple, slender, and not suited for jumping. The abdomen comprises 11 segments, with the cerci long and flexible and the male genitalia asymmetrical.

The ovipositor comprises three pairs of slender, tapering, partly free valves. Although grylloblattids are normally considered to be cool-adapted insects, they cannot withstand temperatures much below 0°C. At –5.5 to 8.0°C they stress. Contrary to the popular belief that they can withstand very low temperatures, they can be killed by ice formation within the body as a result of their low levels of glycerol, sorbitol, or erythrol. Copulation has been observed in a few species. In *G. nipponensis*, the female is chased and seized by the male.

The resulting copulation can last from 30 min to 4 h. Males always assume a position on the right side of the female as a response to the male's asymmetrical genitalia. Oviposition occurs 10 to 50 days after copulation.

Life Cycle

Females lay eggs with the elongate ovipositor in wood or under stones and decaying plant material. None have been found in moss. Oviposition for each egg takes about 3 min, and females lay 5 or 6 eggs per day to a total of about 30 eggs. A captive female laid 145 eggs in her lifetime. The large, black eggs, develop over periods of from 5 months to 3 years.

Habitat

Rock crawlers are only found in the Northern hemisphere with locations ranging from the United States and Canada to Japan and Siberia. They prefer higher elevations, most commonly found between 656 and 10,499 feet. They are usually found in mixed forests, above the point which trees can grow, usually near snowfields.

Homoptera

Common names: cicadas, leafhoppers, aphids, scales, etc. Homopteran (order Homoptera). Homoptera is a deprecated suborder of order Hemiptera; recent morphological studies and DNA analysis strongly suggests that the order is paraphyletic. It was therefore split into the suborders Sternorrhyncha, Auchenorrhyncha, and Coleorrhyncha, although there is currently debate over the monophyly of Auchenorrhyncha. This group includes aphids, cicadas, and leafhoppers. They are all "true bugs," and they all have sucking mouthparts.

More than 32,000 species of sucking insects, the members of which exhibit considerable diversity in body size. All of the Homoptera are plant feeders, with mouthparts adapted for sucking plant sap from a wide assortment of trees and wild and cultivated plants. Many homopterans cause injuries or destruction to plants, including fruit trees and grain crops, and can be vectors of plant diseases. A few provide secretions or other products that are beneficial and have commercial value. Most members of the Homoptera fall into one of two large groups; the Auchenorrhyncha, which consists of the cicadas, treehoppers, froghoppers or spittlebugs, leafhoppers, and planthoppers or fulgorids; and the Sternorrhyncha, which includes aphids or plant lice, phylloxerans, coccids, scales, whiteflies, and mealybugs.

Horntail

Kingdom: Animalia, **Phylum**: Arthropoda, **Class**: Insecta, **Order**: Hymenoptera, **Suborder**: Symphyta, **Superfamily**: Siricoidea, **Family**: Siricidae The pigeon horntail is classified as *Tremex columba.*

Horntail or wood wasp is the common name for any of the 100 non-social species of the family Siricidae, of the order Hymenoptera, a type of xylophagous sawfly. This family was until recently believed to be the sole living representative of the superfamily Siricoidea, a group well-represented in early Tertiary and Mesozoic times, but the family Anaxyelidae has recently been linked to this group. The last

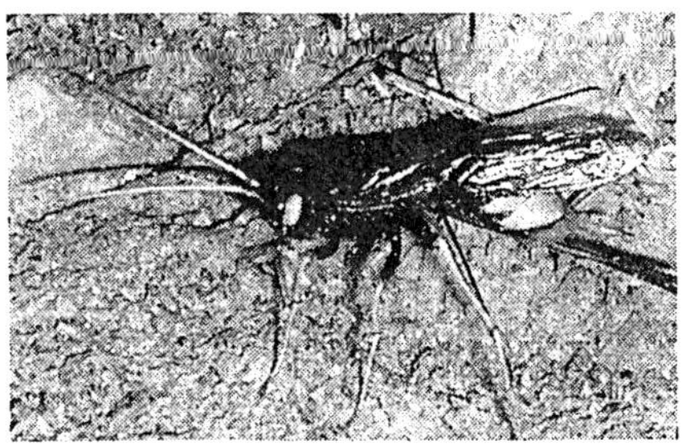

tergite of the abdomen has a strong, projecting spike, thus giving the group its common name (the ovipositor is typically longer and also projects posteriorly, but it is not the source of the name). A typical adult horntail is brown, blue or black with yellow parts in colour, and may often reach up to 4 cm long. The pigeon horntail (Tremex columba) can grow up to 5 cm long (not counting the ovipositor), among the longest of all Hymenoptera.

Female horntails lay their eggs in trees. The larvae bore into the wood and live in the tree for up to 2 years, possibly more. They typically migrate to just under the bark before pupation.

Contrary to the older book quoted below, which refers to a "slight side-to-side motion", the New Zealand wasp pictured here was alternately contracting muscles in its thorax then abdomen to slide the two halves of the ovipositor up and down. Note the spiral groove on the ovipositor, visible on the photograph but not easily to the naked eye.

Life Cycle

Adult horntails are attracted to weakened, dying, or recently cut wood. The female inserts her ovipositor as much as 3/4 inch into the wood to lay her eggs. Eggs are laid singly, but each female is capable of laying 300 to 400 eggs. After 3 to 4 weeks, eggs hatch and the larvae chew their way into the sapwood and heartwood. The tunnels are bored during the 2 to 3 year time span of the larval stage. Tunnels may be from 1 to 2 feet in length. As the larvae chew, the tunnels become packed with a sawdust-like material, making them difficult to see to the uninformed eye. Pupation occurs in the tunnel made by the larva, in a parchment-like cocoon. The adults emerge in August and September, and are often seen resting on a tree stump or freshly cut pile of logs. One interesting aspect of the life of Tremex columba is that there is a spectacular looking ichneumon wasp parasitoid that attacks it. The ichneumon wasp female drills her long ovipositor into the wood where the horntail larva is working. The ichneumon's ovipositor enables it to drill into the wood 3 to 4 or more inches to reach the horntail larva. She lays her eggs either on or near the horntail and the parasitic larva begins to feed. After pupation, the ichneumon chews its way out of the bark and may occasionally be seen resting on a tree. It is conspicuous because of its very long ovipositor (3 or more inches long).

Habitat and Habits

The various species of horntails use their ovipositor for laying eggs in the wood of conifers. The Banded Horntail shows a preference for spruce. To be able to drill through solid wood the horntails need a very powerful and rigid ovipositor. Horntails tend to attack weak, dying or recently dead trees and thus are often found in logged areas or fresh burns. The larvae bore long tunnels usually in the first 4 cm of the wood. They can take 2 to 3 years to mature. In most cases, it will not be necessary to do anything to control horntails.

The adult horntail does not re-infest seasoned wood. It will not lay eggs in wood that is inside the home. Damaged wood can often be repaired or replaced. Firewood should not be stored inside the home. Firewood should only be brought in when it will be used. Sometimes there are a large number of horntails in the same area. In those situations, it may become necessary to treat.

There are special liquid insecticides that are made for treating unfinished lumber. The liquid is designed to kill the larvae and adult insects that are still inside the wood. These products might not be available at the home store. A pest control professional that specializes in termite control should have these products and will be able to do this treatment.

Hymenoptera

Kingdom: Animalia, **Phylum**: Arthropoda, **Class**: Insecta, **Superorder**: Hymenopterida, **Order**: Hymenoptera Linnaeus, 1758

Hymenoptera is one of the largest orders of insects, comprising the sawflies, wasps, bees and ants. There are over 130,000 recognised species, with many more remaining to be described. The name refers to the heavy wings of the insects. The hindwings are connected to the forewings by a series of hooks called hamuli. emales typically have a special ovipositor for inserting eggs into hosts or otherwise inaccessible places. The ovipositor is often modified into a stinger. The young develop through complete metamorphosis — that is, they have a worm-like larval stage and an inactive pupal stage before they mature (See holometabolism).

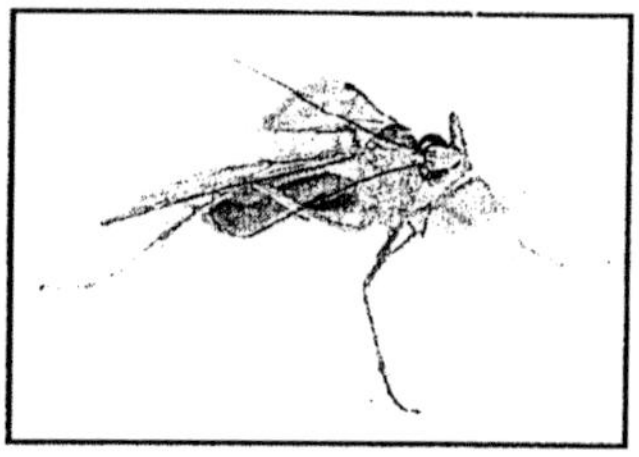

Life Cycles

Life Stages

Hymenoptera undergo complete metamorphosis - that is, they have an egg, larva, pupa and adult stage. The larvae are usually soft and white with no legs, looking similar to diptera larvae; the larvae of some Sawflies have small fore and hind legs like caterpillars. The sexes are separate, but males are rare in several Parasitica, and the sexes are "redefined" a little in the social aculeates. Many Hymenoptera are protandrous, that is the adult male starts to emerge before the female, awaiting the emergence of females to mate with or even mating before or while she is trying to emerge. In contrast, females of some Parasitica will wait for their own male offspring to emerge before promptly mating with them!

Types of Life Cycle

There are a bewildering variety of types of life cycle among Hymenoptera. These are intimately bound up with the way in which food is provided for the offspring. In the Symphyta, eggs are laid in plant tissue and simply abandoned. Other plant-feeding Hymenopterans, such as the Gall Wasps (Parasitica), induce plants to grow protective structures (galls) that shelter the young as they feed. In the Parasitica, and some aculeates, parasitoid life cycles are common. Parasitoid larvae behave like parasites, feeding off the tissues of living animals, but unlike true parasites usually kill the host at the end of their development. Parasitoids are often extremely specialised, attacking only a single species. When this itself is the larva of a parasitoid insect, they may be called hyperparasitoids. Nest-building is well-developed in the aculeates, with social organisation being a special case.

Solitary, as opposed to social, aculeates may build one or more nests, each supplied with food for the larvae. Other Hymenoptera take advantage of the hard work done by others to supply food. This is known as kleptoparasitism, and the species are often called "cuckoo wasps" or "cuckoo bees". Eggs are laid on another species' store of food, which is eaten by the developing larva, usually killing the rightful heir in the process. There are even social kleptoparasites, in which a whole colony of bees, wasps or ants is taken over from within by another, often closely related, species.

Ichneumon

Kingdom: Animalia, **Phylum**: Arthropoda, **Class**: Insecta, **Order**: Hymenoptera, **Suborder**: Apocrita, **Superfamily**: Ichneumonoidea, Latreille, 1802.

The Ichneumonoidea are insects classified in the hymenopteran suborder Apocrita. The superfamily is made up of the ichneumon wasps (often inaccurately called "ichneumon flies"; family Ichneumonidae) and the braconids (family Braconidae). Like other parasitic wasps, they were long placed in the "Parasitica", variously considered as an infraorder or an unranked clade, but actually not a monophyletic group. The superfamily Ichneumonoidea has been estimated to contain well over 80,000 different species. In the largest family of ichneumons the first abdominal segment is large and rounded. A common American species is about 25 mm (1 in) long.

The ichneumon wasps are more familiar to non-entomologists, being larger and about three times as diverse as the braconids. They are solitary insects, and most are parasitoids—the larvae feeding on or in another insect which finally dies. Being in the same order, ichneumons are closely related to other hymenopterans such as ants and bees. Members of the family Ichneumonidae are usually larger than members of the Braconidae, and are distinguished primarily by details of wing venation. Many species in both families utilize polydnaviruses to suppress the immune system of the host insect. Some members use many different insects as a host, others are very specific in host choice. Various ichneumons are used successfully as biological control agents in controlling pests such as flies or beetles. Ichneumon wasp species are highly diverse, ranging from 3 millimetres (0.12 in) to 130 mm (5.1 in) long. Most are slender, with the females of many species (particularly in the genus Megarhyssa) having an extremely long ovipositor for laying eggs.

Life Cycle

The female finds a host and lays an egg on, near, or inside the host's body. Upon hatching, the larval ichneumon feeds either externally or internally, killing the host when they themselves are ready to pupate. Despite looking formidable, the ovipositor does not deliver a sting like many wasps or bees. It can be used by the wasps to bore into and lay eggs inside rotten wood.

Habitat

Ichneumons are over far parts Africa spread and are missing only in dry deserts and much seal to rain forests. Their preferential Habitat is those Savannah. In addition belong Small Asia and Palestine to their natural circulation area. As only Mangustenart they live also in south Europe, in Spain and Portugal. Whether the there existence of natural origin are or were introduced by humans in the antique one, is disputed. In recent past they became also in Italy and up Madagascar introduced.

Katydid

Kingdom: Animalia,**Phylum**: Arthropoda, **Class**: Insecta, **Order**: Orthoptera, **Suborder**: Ensifera, **Infraorder**: Tettigoniidea, **Superfamily**: Tettigonioidea, Krauss, 1902, **Family**: Tettigoniidae, Krauss, 1902

The family Tettigoniidae, known in American English as katydids and in British English as bush-crickets, contains more than 6,400 species. It is part of the suborder Ensifera and the only family in the superfamily Tettigonioidea. They are also known as long-horned grasshoppers, although they are more closely related to crickets than to grasshoppers. Katydid emulating a leaf. Many tettigoniids exhibit mimicry and camouflage, commonly with shapes and colours similar to leaves.

The katydid is named for its song, which has been described as "Katy-did, she-did." The longer "Katy-didn't" is rarer. Scientific literature makes no mention of what exactly it is that Katy did or did not do. The exact song made by the rubbing of wings is distinct between populations across the insect's range.

Life Cycle

Eggs, which the female lays in crevices in bark and on stems, overwinter, and then hatch in the spring. The young feed on surrounding foliage, as do adults, and most likely never leave the tree on which they are born. They begin calling around July, then die in the frosts of October or November.

TRUE KATYDID

The true katydid gets its name because it was the first species to have its call transcribed. They inhabit deciduous forests from Massachusetts to Florida and west to Texas and Kansas and northeast to Ontario. Their antennae is longer and stiffer that other katydids. Females lay one generation of eggs per year in the crevices of bark and soft plant tissue. If jostles or startled, they squawk loudly, raise its forewings, leap from their perch and flutter downward. Once they reach the ground, they walk to the nearest tree and climb the tree trunk.

FALSE KATYDID

The false katydid gets its name from the rapid "tic-tic-tic-tic" sound it makes, unlike the more traditional katydid call. They are short winged, have greenish to brown bodies that grow up to 2 ½ inches long. They have long antennae and hind legs for jumping. Males generate their sound by rubbing their wings together. The false katydid is found throughout the southeast. Depending on the species, some are native to a particular state.

MEADOW KATYDID

Similar to the cone-headed katydid, the meadow katydid has a rounded cone that projects beyond their antennae segments. Their bodies are brown and green and female adults feed at night on the seed of grasses. There are two types of the meadow katydid found in the eastern US.

The Lesser Meadow Katydid has a slender body and wings that don't extend past their abdomen. They also have a straight ovipositor for laying eggs. They are found in Eastern and Central North America and lives in fields and meadows. They feed on leaves, flowers, seed and the pollen of grasses. They produce one generation per year.

The Greater Meadow Katydid is more robust with wings that extend past the tip of their abdomen and a curved ovipositor. They are found in Eastern and Central North America with one species that reached California. They prefer to inhabit areas that are moist.

CONE-HEADED KATYDID

The Cone-headed Katydid has wings that extend beyond their abdomen and a cone shaped head that is separated from its face by a gap. The cone can be straight pointed, bent pointed or round tipped depending on the species and the area they inhabit. They live from the Eastern US and Southeaster Canada and in the Southwestern US. They have oversized jaws and are strong fliers. There are currently 22 species of the Cone-headed katydid located in North America. They are identified by their brown or green colors. In the winter males are brown while the females are green, and during the summer both species are green. The adult female uses her ovipositor to wedge eggs between the stems and sheaths of root leaves or cattail and grasses. Mating usually begins in January. The Cone-headed Katydid is the only known katydid to have different calling songs for each season. Males often succumb to the phonotatic parasitoid fly.

KATYDIDS THAT ARE PESTS

Shield Backed Katydid

There are currently 122 species of the Shield Backed Katydid in North America. They are identified by their short wings, brown or black body and a back that looks like a shield that extends over its wings. The Shield Backed Katydids resemble robust crickets. They are also known as Wart Biters in Europe because they can bite if handled. Most of the shield backed katydids are flightless. These insects are found in the western US with four species found in the East. This Katydid inhabit open country and croplands. Some prey on other insects, while most eat plant material and dead insects. They are related to the Mormon Cricket and can be pests at times by feasting on crops.

Greater Angle-Winged Katydid

The Greater Angle-Winged katydid is a crop damaging pest throughout Florida. They chew on leaves and the fruit of citrus trees. This katydid can grow up to 2 ½ inches long, is bright green and has wings that resemble citrus leaves. They over winter and hatch in the spring bringing forth several generations per year. Their population is usually increased from June to September. The eggs of the Greater Angle-Winged Katydid are laid out in rows along the edge of leaves in large trees.

Greater Angle-Winged Katydid feed on the foliage of citrus trees during mid to late morning and severe defoliation can occur in young trees. Sometimes they feed on the peel of growing oranges like some species of grasshoppers. The results are smooth sunken areas in the rind as the fruit develops.

Habitat

Katydids are found throughout the world, in habitats ranging from tropical to temperate or arid environments. Angular winged katydids are found in temperate and arid regions of the southern United States.

Control

No control measure has been identified.

Lacewing (Chrysopidae)

Kingdom: Animalia, **Phylum**: Arthropoda, **Subphylum**: Hexapoda, **Class**: Insecta, **Subclass**: Pterygota, **Infraclass**: Neoptera, **Superorder**: Endopterygota or Neuropterida, **Order**: Neuroptera, **Suborder**: Hemerobiiformia, **Superfamily**: Chrysopoidea, **Family**: Chrysopidae.

Green lacewings are insects in the large family Chrysopidae of the order Neuroptera. There about 85 genera and (differing between sources) 1,300–2,000 species in this widespread group. Members of the genera Chrysopa and Chrysoperla are very common in North America and Europe; they are very similar and many of their species have been moved from one genus to the other times and again, and in the non-scientific literature assignment to Chrysopa and Chrysoperla can rarely be relied upon. Since they are the most familiar neuropterids to many people, they are often simply called 'lacewings'. But actually most of the diversity of Neuroptera are properly referred to as some sort of 'lacewing', so common lacewings is preferable.

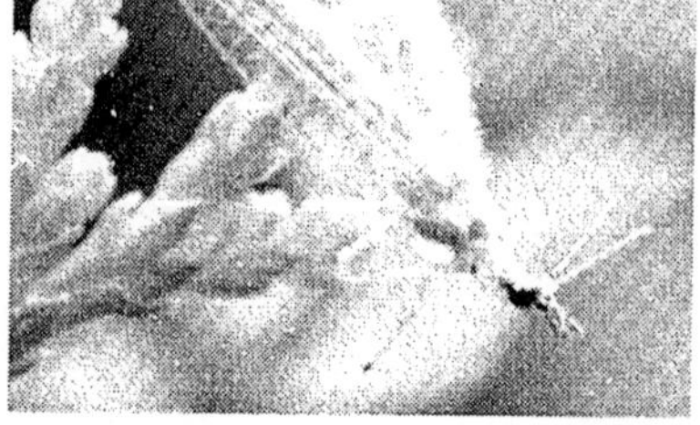

While depending on species and environmental conditions, some green lacewings will eat only about 150 prey items in their entire life, in other cases 100 aphids will be eaten in a single week. Thus, in several countries, millions of such voracious Chrysopidae are reared for sale as biological control agents of insect and mite pests in agriculture and gardens. They are distributed as eggs, since as noted above they are highly aggressive and cannibalistic in confined quarters; the eggs hatch in the field.

Their performance is variable; thus, there is a lot of interest in further research to improve the use of green lacewings as biological pest control. Species that have hitherto attracted wider study and are more or less readily available as captive-bred eggs to deposit out for hatching in pest-infested plant cultures are several members of Chrysoperla as well as Mallada signatus. Gardeners can attract these lacewings – and therefore ensure a steady supply of larvae – by using certain companion plants and tolerating beneficial weeds. Chrysopidae are attracted mainly by Asteraceae – e.g. calliopsis (*Coreopsis*), cosmos (*Cosmos*), sunflowers (*Helianthus*) and dandelion (Taraxacum) – and Apiaceae such as dill (*Anethum*) or angelica (*Angelica*).

Life Cycle

The larvae of most families are predators. Many chrysopids eat aphids and other pest insects, and have been used for biological control (either from commercial distributors but also abundant and widespread in nature). Larvae in various families cover themselves in debris (sometimes including dead prey insects) as camouflage, taken to an extreme in the ant lions, which bury themselves completely out of sight and ambush prey from 'pits' in the soil. Larvae of some Ithonidae are root feeders, and larvae of Sisyridae are aquatic, and feed on freshwater sponges. A few mantispids are parasites of spider egg sacs.

As in other holometabolic orders, there is a pupal stage, generally enclosed in some form of cocoon composed of silk and soil or other debris. The pupa eventually cuts its way out of the cocoon with its mandibles, and may even move about for a short while before undergoing the moult to the adult form.[4]

Adults of many groups are also predatory, but some do not feed, or consume only nectar.

Habitat

Adults are often found on vegetation. The larvae are usually more specific in their selection of habitats, preferring certain soil types, freshwater habitats, or other locations that guarantee the availability of certain kinds of prey. The larvae of antlions and spoonwing lacewings prefer sandy habitats mostly in drier regions. Those of green and brown lacewings are found only on shrubs and trees. Larval dustywings also prefer trees and shrubs and are usually quite specific about the species of plant. Others prefer freshwater streams or hunt under rocks or in leaf litter along the shore. The larvae of moth lacewings are grublike and live among the roots of plants.

Control

Control is easiest when lacewings are detected early. Acephate is the active ingredient in several insecticides that are readily available for lacewing control. Remember, however, that once the leaves become ugly and discolored, they will not recover. Those leaves are damaged, and the focus should be to stop the problem before the lacewings spread to more leaves or more azaleas.

Leaf Insect (Phylliidae)

Kingdom: Animalia, **Phylum**: Arthropoda, **Class**: Insecta, **Order**: Phasmatodea, **Superfamily**: Phyllioidea, **Family**: Phylliidae, Redtenbacher, 1906.

The family Phylliidae (often misspelled Phyllidae) contains the extant true leaf insects or walking leaves, which include some of the most remarkable leaf mimics in the entire animal kingdom. They occur from South Asia through Southeast Asia to Australia. At present, there is no consensus as to the preferred classification of this group; some sources treat Phylliidae as a much larger taxon, containing the members of what are presently considered to be several different families. Leaf insects use camouflage to take on the appearance of a leaf.

They do this so accurately that predators often aren't able to distinguish them from real leaves. In some species the edge of the leaf insect's body even has the appearance of bite marks. To further confuse predators, when the leaf insect walks, it rocks back and forth, to mimic a real leaf being blown by the wind. The scholar Antonio Pigafetta may have been the first to document the creature. Sailing with Ferdinand Magellan's circumnavigational expedition, he studied and chronicled the fauna on the island of Cimbonbon as the fleet hauled ashore for repairs.

Life Cycle

Male leaf insects have thinner bodies, larger wings and longer antennae. When they are ready to mate, males fly around to locate females. Female leaf insects lay eggs throughout most of their adulthood. Depending on the particular species, female leaf insects lay up to 1,000 eggs over their lifetime, dropping them onto the ground.

Egg

Depending on the species, eggs might stay on the ground until they hatch. In the case of spiny leaf insects, the eggs have special parts that are attractive to ants and encourage ants to carry the eggs back to their nest where the eggs are safe from predators. In captivity, leaf insect eggs are placed in a plastic box for incubation and are sprayed lightly every day. After four to six months, the eggs hatch. Warmer temperatures help the eggs to hatch sooner.

Nymph

When it emerges from the egg, a leaf insect is in the nymph form. Small and brown, it looks like an adult leaf insect without the wings. It feeds on young, soft leaves after a few days. It needs humidity and clean surroundings to survive, and often mortality rates can be as high as 20 percent. In captivity, you can place several leaf insect nymphs in one cage to increase appetite. Unlike mammals, leaf insects do not have skin that grows as they become older. Instead, nymphs shed their skin several times to reveal bigger skin underneath, a process also known as molting. Depending on the particular species, a leaf insect might take weeks or months to reach adulthood.

Adult

At adulthood, female leaf insects tend to be bigger than males. For example, adult Javan leaf insect males range between 1.8 to 2.7 inches, while females often measure between 2.6 to 3.7 inches in length. When a leaf insect reaches adulthood, it begins to eat both young and old leaves. Leaf insects have a life expectancy of about one or two years, with females often outliving males, according to LeafInsects.net.

Habitat

Stick insects are found over much of Australia where adequate trees and shrubs occur to feed on. Those species known strictly as leaf insects are only found in northern Queensland. *Eurycnema goliath* is a conspicuous, heavily built species known as the Goliath stick insect and can be found over much of northern Australia and down the east coast as far as Victoria. Females of this species are fully winged and known to grow up to 25 centimetres in length.

Leafhopper

Kingdom: Animalia, **Phylum**: Arthropoda, **Class**: Insecta, **Order**: Hemiptera, **Suborder**: Clypeorrhyncha, **Superfamily**: Membracoidea ,**Family**: Cicadellidae, Latreille, 1802.

Leafhopper is a common name applied to any species from the family Cicadellidae. Leafhoppers, colloquially known as hoppers, are minute plant-feeding insects in the superfamily Membracoidea in the order Homoptera. They belong to a lineage traditionally treated as infraorder Cicadomorpha in the suborder Auchenorrhyncha, but as the latter taxon is probably not monophyletic, many modern authors prefer to abolish the Auchenorrhyncha and elevate the cicadomorphs to a suborder Clypeorrhyncha. Leafhoppers are found all over the world, and constitute the second-largest hemipteran family. They have at least 20,000 described species. The tribe Proconiini of the subfamily Cicadellinae is commonly known as sharpshooters.

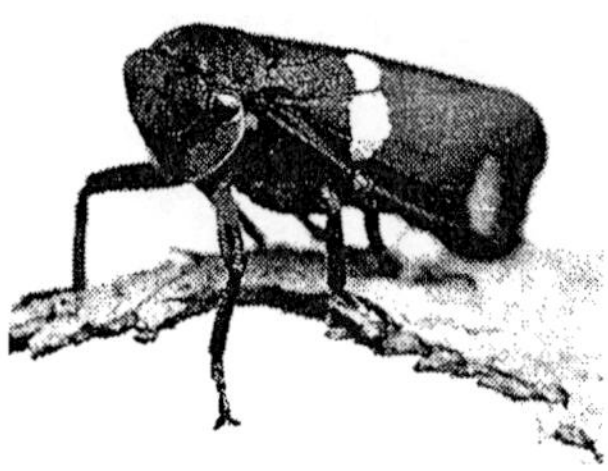

Many species of leafhoppers exist in home gardens throughout North America. Both adults and nymphs feed by puncturing the undersides of leaves and sucking out plant juices. Their toxic saliva causes spotting (white specks), yellowing, leaf curling, stunting and distortion of plants. They are also responsible for transmitting the organisms causing virus diseases in plants. Common host plants include beans, lettuce, beets, potato, grapes, roses, and many others.

Life Cycle

Most species overwinter as eggs or adults. Eggs are inserted into leaf veins, shoots or stems of host plants. Wingless nymphs hatch from eggs in about 10 days and begin feeding on the tender new growth of their host plant. They develop through five stages (instars) over a period of 12 to 30 days, leaving shed skins in the feeding area. As nymphs grow larger, they develop wing pads. Most leafhoppers produce one generation per year, but some may develop up to six.

Habitat

Species can be somewhat specific to certain host plants. As a group they feed on leaves of a wide variety of plants including many types grasses, flowers, vegetables, fruit trees, shrubs, deciduous trees, palms and weeds. The rose leafhopper, Edwardsiana rosae (Linnaeus), feeds primarily on plants of the rose family, although foliage of other woody plants (blackberry, Cornus, oak, Prunus, Populus, raspberry, Ulmus, Acer and others) serve as food. The potato leafhopper, Empoasca fabae (Harris), feeds on legumaceous plants like alfalfa, as well as on apple, birch, chestnut, maples, and others. Species in the genus, Erythroneura, feed on sycamore leaves, but also on apple, grape and willow. The aster or six-spotted leafhopper, Macrosteles quadrilineatus Forbes, feeds on vegetables and annual flowers and spreads the aster yellows virus to woody plants like periwinkle and Thunbergia species. Nymphs and adults feed on the underside of leaves. Some leafhoppers are readily attracted to lights.

Control

Remove garden trash and other debris shortly after harvest to reduce over-wintering sites. Floating row covers can be used as a physical barrier to keep leafhoppers from damaging plants. Commercially available beneficial insects, such as ladybugs, lacewing, and minute pirate bugs, are all voracious predators of both the egg and young larval stage. Apply diatomaceous earth to plants and/or spot treat with insecticidal soap to keep pest populations under control. Thorough coverage of both upper and lower infested leaves is necessary for effective control. If pest levels become intolerable, spot treat with botanical insecticides as a last resort.

Leafminer

Class: Insecta, **Order**: Lepidoptera, **Family**: Lyonitidae.

Leaf miner is a term used to describe the larvae of many different species of insect which live in and eat the leaf tissue of plants. The vast majority of leaf-mining insects are moths (Lepidoptera), sawflies (Symphyta) and flies (Diptera), though some beetles and wasps also exhibit this behaviour. Like Woodboring beetles, leaf miners are protected from many predators and plant defenses by feeding within the tissues of the leaves themselves, selectively eating only the layers that have the least amount of cellulose. When attacking Quercus robur (English oak), they also selectively feed on tissues containing lower levels of tannin, a deterrent chemical produced in great abundance by the tree. The precise pattern formed by the feeding tunnel is very often diagnostic for which kind of insect is responsible, sometimes even to genus level.

The mine often contains frass, or droppings, and the pattern of frass deposition, mine shape and host plant identity are useful to determine the species of leaf miner. A few mining insects utilise other parts of a plant, such as the surface of a fruit. Some patterns of leaf variegation are part of a defense strategy employed by plants to deceive adult leaf miners into thinking that the leaf has already been predated.

Life Cycle

Egg. The eggs are oval, flattened against the leaf surface; translucent, greenish white with granulate surface which turns yellowish when about to hatch.

Larva. The emerging larvae are distinctly segmented with a rather pointed heads and abdomens. A mature larva measures 5.5 mm long. The larva has a yellowish body with paired pink spots on the dorsolateral sides of the thorax which later disappear and are replaced by red tubercles in all segments.

Pupa. The pupae measuring 3.5 mm appear green at first with mottled red markings. Later the red markings disappear and they turn dark brown with lateral projections on the abdomen.

Adult. The adults are very small moths, 3.5 - 4.0 mm long with grayish to brown bodies and light brown scales.

Habitat

Leafminers can be found in garden foliage as well as the leaves of birches, arborvitae and junipers. These larvae are found throughout North America.

Control

Biological control: Leaf miners are generally controlled by predators and parasites like Apanteles sp.

Chemical control: The insecticides recommended for leaf miner control include carbaryl, chlorfenvinphos, diazinon, dimethoate and trichlorphon.

Locust

Kingdom: Animalia, **Phylum**: Arthropoda, **Superclass**: Hexapoda, **Class**: Insecta, **Order**: Orthoptera, **Family**: Acrididae, **Subfamily**: Cyrtacanthacridinae, Oedipodinae, Gomphocerinae.

Locust is the swarming phase of short-horned grasshoppers of the family Acrididae. These are species that can breed rapidly under suitable conditions and subsequently become gregarious and migratory. They form bands as nymphs and swarms as adults—both of which can travel great distances, rapidly stripping fields and greatly damaging crops. The origin and apparent extinction of certain species of locust—some of which reached 6 inches (15 cm) in length—are unclear. Locusts are an edible insect and are considered a delicacy in some countries and in history.

Life Cycle

Female locusts prefer to lay eggs in hard soil and often congregate on tar roads trying to drill into the tar to lay eggs. Female locusts lay eggs in batches called pods, usually at a depth of 2–10 cm. Each pod will contain 30–60 pale yellow banana shaped eggs 5–6 mm long and an individual female may lay up to three pods. A collection of egg pods laid by a number of locusts is termed an egg bed.

Eggs need warmth and moisture to develop and will suspend development if these needs are not met. In summer, eggs may hatch within 14–16 days. An immature locust is called a nymph or hopper. After hatching from the egg, a locust goes through five growth stages called instars, moulting at each stage.

At each stage, the developing wings become more noticeable until the locust becomes a fledgling adult and then a mature adult capable of sustained flight. Normally, the nymphs take 4–8 weeks to complete this development.

When the population is large and at the second and third instar, nymphs will often concentrate into dense aggregations called bands. These bands will vary in size but can extend over several kilometres and can move over 1 km from the egg bed before the nymphs fledge and can consume or damage all vegetation in its path.

After the final moult, the adult locust emerges with fully-formed wings. At first the body and wings of the locust are still soft and, until they harden after about a week, sustained flight is not possible. Green feed is required to provide fuel for flight and egg development. If numbers are sufficient, adult locusts may concentrate into dense groups called swarms. These infest areas usually less than 5 Km2 but which can be up to 50 Km2. If sufficient green feed is available to enable flight but conditions are drying off, migration of locusts may occur.

Habitat

The locust's usual habitat stretches from West Africa to India and from Spain to Tanzania. In most of these countries, the population depends mainly on agriculture. However, the first swarms are often hatched in the Sahel, the huge band of arid land just south of the Sahara, from Senegal to Egypt.

Control

Control by spreading poisoned food among the bands is very effective, but it is cheaper to spray concentrated insecticide solutions from aircraft over the insects or the vegetation on which they feed. They eat the equivalent of their own weight in a day, and, flying at night with the wind, may cover some 500 km/300 mi. The largest known swarm covered 1,036 sq km/ 400 sq mi, comprising approximately 40 billion insects.

A biological pesticide to control locusts was being trialled across Africa by a multinational team in 1997. Dried fungal spores sprayed in breeding areas pierce the locust exoskeleton on germination and invade the body cavity causing death. The fungus is passed from insect to insect and persists in the area making repeated treatments unnecessary.

Louse

Kingdom: Animalia, **Phylum**: Arthropoda, **Class**: Insecta, **Subclass**: Pterygota, **Infraclass**: Neoptera, **Order**: Phthiraptera, Haeckel, 1896.

Lice (singular: louse) is the common name for over 3000 species of wingless insects of the order Phthiraptera; three of which are classified as human disease agents. They are obligate ectoparasites of every avian and mammalian order except for Monotremes (the platypus and echidnas), bats, whales, dolphins, porpoises and pangolins. Body Lice, *Pediculus humanus humanus*, Head Lice, *Pediculus humanus capitus.*

Life Cycle

They have a three-stage life cycle: egg, nymph and adult. The eggs of lice are called nits. They are light tan colored, oval cylinders. The eggs of head lice are glued to hairs of the head. The favorite areas seem to be near the ears and back of the head. Body lice glue their eggs to clothing, especially near seams and creases. Female head lice lay 6--7 eggs per day and may lay a total of 50--150 eggs during their lives. Female body lice may lay a total of 270--300 eggs in a lifetime. Under normal conditions the eggs will hatch in about a week. Below 74 degrees F, most eggs will not hatch. Newly hatched nymphs must feed within 24 hours or die. There are three immature nymphal stages separated by molts. Their appearance and habits are not greatly different from those of adults. Both nymphs and adults have piercing-sucking mouthparts, which pierce the skin for a blood meal. Adults may survive 3 to 5 days without a blood meal. Normally a young louse will mature to an adult in 3 to 5 weeks.

Habitat

Head lice prefer to live on the hair of the head although they have been known to wander to other parts of the body. Body lice tend to aggregate where clothing most frequently touches the skin. Body louse nymphs spend the greater part of their time in the clothing, and feeding on the host only takes place when the host is resting or sleeping. All stages of the body louse congregate together, being attracted to each other by smell, e.g., the odor of the excrement. Body lice crawl about the clothing, generally keeping close to the host's body, although in heavy infestations they may be seen crawling outside on the outer garments. Lice are most often transmitted from person to person by direct contact, and they can be easily dislodged from a lousy person onto an uninfested person. They are most often a problem in unsanitary conditions. Simply washing clothes regularly can reduce their abundance.

Control

Non-chemical Management: An alternative that has become popular recently is to attempt to suffocate head lice by coating hair overnight with vegetable oil, mayonnaise, baby or mineral oil, or hair gel. Although there are anecdotal reports of this method being successful, there are also reports of failure as well. A drawback to this method is that removing this material from the hair is usually difficult. This method has not been scientifically tested and care should be taken if using this method. Physical removal is an important tactic in head lice control. Use nit combs or cat flea combs to remove the lice and eggs. Because eggs can hatch for up to 10 days after they are laid, comb hair daily for two weeks after you find the last live head lice, to remove any young that have emerged. Nits found further than 1/2 inch from the scalp are probably not viable and can be ignored.

You can purchase nit combs through pharmacies or online on the internet. All bedding and clothing of infested people should be washed and dried at a high temperature (150°F or higher). Pillows or other nonwashable items should be placed in plastic bags and placed in a freezer for several days. It is not necessary to clean other areas of schools or homes. Head lice that are dislodged survive for just a short time off a host. Combing for nits and cleaning are important even if you choose to use chemical management.

Insecticidal Management: There are several insecticidal products available to control head lice. Insecticides containing permethrin and synergized pyrethrins are common in shampoos and lotions than can be purchased over-the-counter. If you use one of these products, follow their directions precisely, especially the amount of product to use, frequency of application, and whether the hair should be wet! Remember, these products contain insecticides and should be used with great care and respect. Most products recommend reapplying the product in 7 to 10 days after the first application to kill newly hatched lice. Lindane is available by prescription. However, this is a more toxic product and its use is discouraged. Head lice maintain an intimate relationship with their host and cannot survive off a human for more than two days. Because of this, control should concentrate on treating infested individuals. Never treat furniture, bedding, floor or walls inside a school or home with insecticide to control head lice. Physical removal of lice (nit picking) together with chemical application will increase treatment success.

Mantis

Kingdom: Animalia, **Phylum**: Arthropoda, **Class**: Insecta, **Subclass**: Pterygota, **Infraclass**: Neoptera, **Superorder**: Dictyoptera, **Order**: Mantodea, Burmeister, 1838, Carolina mantis is classified as *Stagmonantis carolina*.

Mantodea or mantises is an order of insects that contains approximately 2,200 species in nine families worldwide in temperate and tropical habitats. Most of the species are in the family Mantidae. Historically, the term "mantid" was used to refer to any member of the order because for most of the past century, only one family was recognized within the order; technically, however, the term only refers to this one family, meaning the species in the other eight recently established families are not mantids, by definition (i.e., they are empusids, or hymenopodids, etc.), and the term "mantises" should be used when referring to the entire order.

A colloquial name for the order is "praying mantises", because of the typical "prayer-like" stance, although the term is often misspelled as "preying mantis" since mantises are predatory. In Europe, the name "praying mantis" refers to Mantis religiosa. The closest relatives of mantises are the orders Isoptera (termites) and Blattodea (cockroaches), and these three groups together are sometimes ranked as an order rather than a superorder. They are sometimes confused with phasmids (stick/leaf insects) and other elongated insects such as grasshoppers and crickets.

Life Cycle

Generally, the praying mantis will live for only six months. Some species will live up to a whole year. The life cycle of the praying mantis consists of three main stages. These are the egg stage, nymph stage and adult stage.

The egg stage takes place right before winter. During the autumn season after fertilization the adult female praying mantis will lay 100 to 400 eggs. She begins to secure the eggs on a safe stem or sturdy leave with a frothy liquid that hardens and becomes a protective casing for the eggs during the winter. It is during this life cycle that the praying mantis' eggs are sought after, usually for business, farming or scientific study. These egg structures are known as ootheca. For each egg in the ootheca, there is a compartment-like space. To assist their hatching, these compartments contain small one way valve-like structures. The baby mantids break out of the casing during the spring as the warmth naturally expedites the hatching.

The nymph stage blossoms in the spring. After these hungry nymphs get a taste of the outside air for the first time, they will hang around the egg case for a little while. As with spider-lings, some of these baby mantids will try to eat each other during this time. After dispatching, they start hunting for small insects like fruit flies. During this life cycle the tiny mantids are particularly vulnerable to other larger predators and as with many insects, not all the nymphs will survive this stage. Also, the frail nymphs during this stage will be unable to tackle on prey larger and stronger than themselves and will need to learn to be selective when preying.

The adult stage thrives in the summer. The adolescent praying mantis will spend the late spring and early summer months hunting and eating anything that will not eat it first.

They grow at a steady pace, shedding their exoskeleton as much as ten times in a process known as molting, until they reach their full size by the end of summer. Right before each molt, the praying mantis will often become sluggish and will not eat for sometime. They usually select a solid branch to secure their footing to assist with the difficult process. It is at this critical time during each molting process that the praying mantis can be particularly vulnerable to larger predators as each shed can take a few hours. Tragically, some can even get stuck in the old skin during this process and perish. However, when they come out successfully from the molting process, they soon regain their strength and appetite and are bigger and stronger than before.

Habitat

The praying mantis habitat is difficult to define specifically. In the first place, there are a couple of thousand different species of praying mantis. Some are quite large, attaining a length of nearly 6 inches. At the other extreme, some species are barely a centimeter, less than half an inch, in length. All species share several characteristics. They are skilful hunters, and for the most part very patient ones. They can be cannibalistic. They will eat almost anything that isn't too big that gets too close.

Control

Organic gardeners who avoid pesticides may encourage mantises as a form of biological pest control. Tens of thousands of mantis egg cases are sold each year in some garden stores for this purpose. During fall, praying mantis females deposit a sticky egg case on the underside of a leaf or on a twig. If the egg case survives winter, the offspring, called nymphs, emerge in late spring or early summer. The nymphs have voracious appetites and typically cannibalize each other if they don't have an adequate supply of aphids and other small insects. Egg cases are commercially available for placement in landscaping. However, mantises prey on neutral and beneficial insects as well, basically eating anything they can successfully capture and devour.

Mayfly

Kingdom: Animalia,**Phylum**: Arthropoda, **Class**: Insecta, **Subclass**: Pterygota, **Division**: Palaeoptera (disputed), **Superorder**: Ephemeropteroidea, Rohdendorf, 1968, **Order**: 'Ephemeroptera', Hyatt & Arms, 1891.

Mayflies (Order Ephemeroptera) are delicate seasonal insects appearing in late spring and early summer. The adult stage is often mistaken for mosquitoes because of their size and association with either standing or running water. Unlike mosquitoes, mayfly adults do not bite, lacking functional mouth parts, they are also unable to feed. They live only a few hours or days at the most. Some species appear spontaneously in huge numbers and just as quickly disappear.

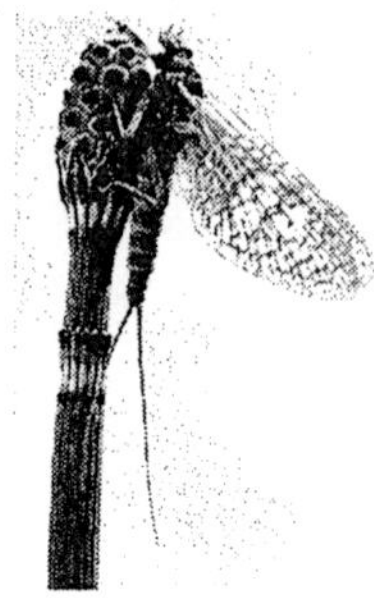

Mayflies are distinctive insects easily recognized by their elongate wings held upright over the body, typical long forelegs (front legs), and characteristic long paired (some species with three) tail filaments extending rearward from the tip of the abdomen. Mayflies vary greatly in length from small species less than ¼ inch in length to "giants" that exceed 2 inches. Colors vary greatly among mayfly species, but most are colored in tans, browns, yellows and greens with dark markings. The prominent forewings (front pair) are usually clear or tinted with an amber to smoky color. An additional feature of mayflies is their large eyes that are subdivided in a large top and small side pairs.

The lifespan of an adult mayfly can vary from just 30 minutes to one day depending on the species. The primary function of the adult is reproduction; the mouthparts are vestigial, and the digestive system is filled with air. The wings are membranous (similar to a house fly's wings but with many more veins) and are held upright like those of a butterfly. The forewings are much larger than the hind wings. In most species, the males' eyes are usually large and the front legs unusually long, for use in locating and grasping females during mid-air mating. In some species, all legs aside from the males' front legs are useless.

Life Cycle

Mayflies are considered aquatic insects with the immature stages, nymphs (naiads), developing in either still or running water. Most nymphs are easily recognized by the presence of usually 3 tail (caudal) filaments, a single tarsal claw, and gill plates on the sides of the abdominal segments. Their diet consists largely of algae, detritis, and other available organic food materials scraped from the bottom, including the surfaces of rocks, gravel, sand grains and rooted aquatic vegetation. Hatching from eggs during the winter, nymphs slowly develop and molt many times until they reach maturity in the spring.

Habitat

Mayflies require clean, unpolluted water that is generally well oxygenated. They are one of the first aquatic species to disappear when water pollution occurs. The use of gasoline motors on our lake and the resulting pollution keeps the Mayfly population on most of our lakes relatively low. Mayflies live in flowing or still water and occasionally can be found at depths to 200 feet. The nymph feeds on algae, diatoms, and vegetable or organic matter that have settled to the bottom. Since their food is mostly on the bottom the larva are usually found on the bottom. They tend to burrow into the bottom substrates and hide under rocks and vegetation.

Control

These insects are completely harmless and, in time, their numbers will decrease in a matter of days to the point where they will become hard to find. Because of their "here today and gone tomorrow" passage in nature, there is no real need to control the adults with pesticides. Many times, the adults will die of natural causes rather than from residual action of pesticides applied to the surfaces of walls and landscape vegetation.

Mecoptera (Scorpion Flies)

Kingdom: Animalia, **Phylum**: Arthropoda, **Class**: Insecta, **Superorder**: Panorpida, (**unranked**): Antliophora, **Order**: Mecoptera, Hyatt & Arms, 1891.

Mecoptera (from the Greek: meco- = "long", -ptera = "wings") are an order of insects with about 550 species in nine families worldwide. Mecoptera are sometimes called scorpionflies after their largest family, Panorpidae, in which the males have enlarged genitals that look similar to the stinger of a scorpion. Soft-bodied insects in which the head is prolonged downwards into a beak with the biting jaws at the lower end. The head bears long, thread-like antennae and large eyes. Two similar pairs of wings are usually present, but a few species are wingless or have atrophied wings. Both sexes have tiny cerci at the tip of the abdomen. The male abdomen is frequently turned upwards and swollen, reminiscent of the tail and sting of a scorpion - hence the common name of the group. Despite their name and appearance, however, the insects are quite harmless. Metamorphosis is complex, with up to seven larval stages and a pupal stage. The larvae resemble caterpillars in general appearance. This is a small Order with only about 300 species, but like the Neuroptera, it is an ancient group. In fact the oldest known fossil endopterygotes - about 250 million years old - are Mecoptera and it is believed that the butterflies and moths,

caddis flies, fleas and true flies all evolved from mecopteran-like ancestors. Some of the present-day Australian Scorpion Flies appear to have survived with little change since Permian times and qualify as 'living fossils'. Four species occur in Britain. While modern mecoptera are overwhelmingly predators or consumers of dead organisms, early ones might have played an important role before the evolution of other insects in pollinating extinct gymnosperms.

Life Cycle

The mecoptera male attracts females with a chemical scent or pheromone. He presents the female a food gift of a dead insect or a salivary mass and mates with her as she is eating. The food gift needs to be large enough to ensure sufficient time to mate, which can take up to 20 minutes and requires at least 5 minutes for adequate transfer of sperm. If the female finds the gift unsatisfactory, she may fly off seeking another male. Males will often try to steal insect food gifts from other males.

Egg: The unusual egg of the Mecoptera expands with development of the embryo. Development can take as little as 5 to 10 days (as with Panorpa and Neopanorpa) or as long as several months (Bittacus stigmaterus). (Ross, 1982)

Larva: The larvae live in damp, shaded areas, feeding on dead or decoying animal matter. Some species are believed to be cannibalistic, with larger individuals feeding on smaller individuals. They are fully grown in approximately 32 days. After 4 instars (stages between molts), the larva burrows a hole in the ground in which to pupate. (Ross, 1982)

Pupa: The prepupal stage can last one to several months, depending on the species. Pupation occurs in the soil. Because the pupa has non-articulate mandibles, it must wait for the cuticle to shed, then uses its mandibles or legs to escape. The adult leaves the cell after 10 to 50 days. (Gullan and Cranston, 2000)

Adult: Adult Mecoptera are mostly carnivorous, feeding on small insects, but will also feed on petals, nectar, fruits, pollen, leaves, and mosses.

Panorpa will feed on insects trapped in spider webs and also on decaying animal matter.

Habitat

Broad-leaved, herbaceous plants shaded by trees are the usual habitat of panorpids, bittacids, and most species in the smaller families. Some species, however, are most commonly found in vegetation near forest borders, whereas others occur in somewhat similar but more uniformly shaded habitats well within a forest.

Elevation clearly affects the distribution of Mecoptera. Boreidae, for example, reach the adult stage in the cold part of the year (and being darkly colored are most often seen on the surface of snow) and at high elevations and high latitudes have a longer seasonal occurrence than boreids at lower elevations. In Mexico, Bittacidae are usually found at elevations below 1500 m, while Panorpidae occur above that level and up to over 3000 m.

Neuroptera

Kingdom: Animalia, **Phylum**: Arthropoda, **Class**: Insecta, **Subclass**: Pterygota, **Infraclass**: Neoptera, **Superorder**: Endopterygota or Neuropterida, **Order**: Neuroptera, Linnaeus, 1758

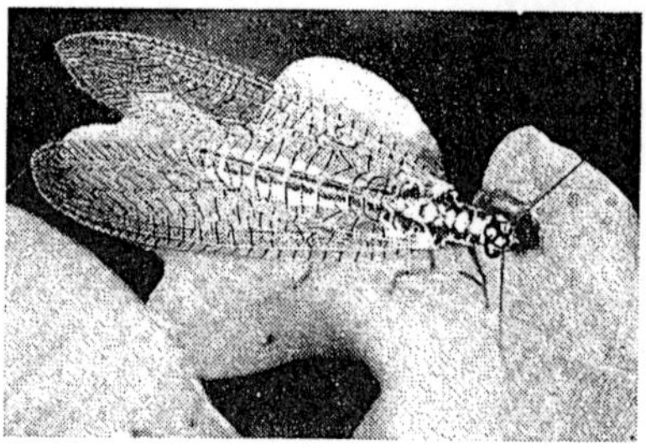

The name Neuroptera is derived from the Greek word 'neuron' meaning sinew and 'ptera' meaning wings. The modern English translation 'nerve-wings' is appropriate because it alludes to the extensive branching found in the wing veins of most Neuroptera.

Life Cycle

The order Neuroptera includes the lacewings and antlions (suborder Planipennia), dobsonflies and alderflies (suborder Megaloptera) and snakeflies (suborder Raphidoidea). "Splitters" prefer to assign each of these groups to a separate order (Neuroptera, Megaloptera, and Raphidioptera, respectively), based on differences in structure and development. The megaloptera are always aquatic as immatures. They live under stones or submerged vegetation and feed on a variety of small aquatic organisms.

Large species, often called hellgrammites, may require several years of growth to reach maturity. Adults usually remain near water, although they are attracted to lights at night. In most species, the adults live only a few days and rarely feed. Except for larval spongillaflies (family Sisyridae) which feed on fresh-water sponges, all members of the suborders Planipennia and Raphidoidea are terrestrial. Antlion larvae live in the soil and construct pitfall traps to snare prey. Lacewing larvae are usually found in vegetation where they typically feed on aphids, mites, and scale insects. Snakefly larvae live in leaf litter or under bark and catch aphids or other soft-bodied prey. In most cases, the adults of these insects are also predators — the non-predatory species usually feed on nectar, pollen, or honeydew. The larvae of antlions and lacewings have specialized mouthparts with large, sickle-shaped mandibles and maxillae that interlock to form pincers. Once impaled on these pincers, a prey's body contents are sucked out through hollow food channels running between the adjacent surfaces of the mandibles and maxillae.

Habitat

Adults are often found on vegetation. The larvae are usually more specific in their selection of habitats, preferring certain soil types, freshwater habitats, or other locations that guarantee the availability of certain kinds of prey. The larvae of antlions and spoonwing lacewings prefer sandy habitats mostly in drier regions. Those of green and brown lacewings are found only on shrubs and trees. Larval dustywings also prefer trees and shrubs and are usually quite specific about the species of plant. Others prefer freshwater streams or hunt under rocks or in leaf litter along the shore. The larvae of moth lacewings are grublike and live among the roots of plants.

Orthoptera

Kingdom: Animalia, **Phylum**: Arthropoda, **Class**: Insecta, **Superorder**: Orthopterida, (**unranked**): Panorthoptera, **Order**: Orthoptera, Latreille, 1793.

Orthoptera is an order of insects with paurometabolous or incomplete metamorphosis, including the grasshoppers, crickets and locusts. Many insects in this order produce sound (known as a 'stridulation') by rubbing their wings against each other or their legs, the wings or legs containing rows of corrugated bumps. The tympanum or ear is located in the front tibia in crickets, mole crickets, and katydids, and on the first abdominal segment in the grasshoppers and locusts.[1] These organisms use vibrations to locate other individuals. Grasshoppers are able to fold their wings, placing them in the group Neoptera.

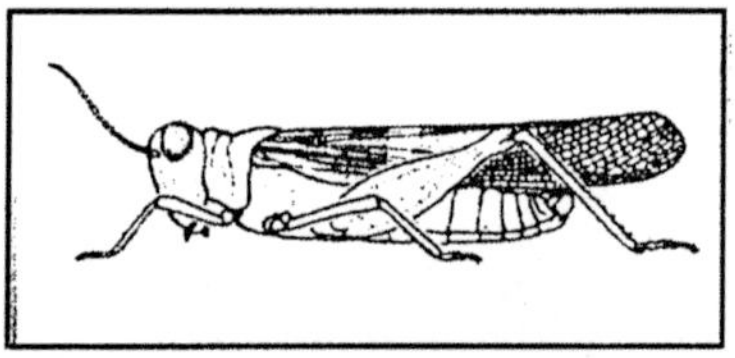

Life Cycle

Orthopteroid species have a paurometabolous life cycle or incomplete metamorphosis. The use of sound is generally crucial in courtship, and most species have distinct songs.[2] Most grasshoppers lay their eggs in the ground or on vegetation. The eggs hatch and the young nymphs resemble adults but lack wings and at this stage are often called hoppers. They may often also have a radically different coloration from the adults. Through successive moults the nymphs develop wings until their final moult into a mature adult with fully developed wings. The number of moults varies between species; growth is also very variable and may take a few weeks to some months depending on food availability and weather conditions.

Habitat

Grasshoppers, crickets, and katydids inhabit virtually all terrestrial habitats, from rocky coastlines, underground burrows, and caves, to the tops of trees in rainforests and high mountain peaks. Others prefer wetlands, with some species living right on the shores of ponds, streams, lakes, and rivers. A few species actually live in water. Many species live in meadows and deserts. Still other species are very particular about where they live; some specialists take shelter in ant nests, while others are at home in greenhouses and basements.

Importance

Orthoptera is generally regarded as a dominant group in most terrestrial habitats. These insects feed on all types of plants and often cause serious economic damage. Swarms of grasshoppers (locusts) regularly appear in parts of Africa, Asia, and North America and destroy crops over wide land areas. Mole crickets are major pests in lawns and golf courses in the southern United States. Several species of field crickets are reared commercially as fish bait.

Phylloxera

Kingdom: Animalia, **Phylum**: Arthropoda, **Class**: Insecta **Order**: Hemiptera, **Suborder**: Sternorrhyncha (was Homoptera), **Superfamily**: Phylloxeroidea, **Family**: Phylloxeridae, **Genus**: Dactylosphaera, **Species**: vitifoliae.

Grape phylloxera (*Daktulosphaera vitifoliae* Fitch (1855) , family Phylloxeridae); originally described in France as Phylloxera vastatrix; equated to the previously described Daktulosphaira vitifoliae, Phylloxera vitifoliae; commonly just called phylloxera is a pest of commercial grapevines world-wide, originally native to eastern North America. These almost microscopic, pale yellow sap-sucking insects, related to aphids, feed on the roots and leaves of grapevines depending on the phylloxera genetic strain. On *Vitis vinifera* L., the resulting deformations on roots (nodosities and tuberosities) and secondary fungal infections can girdle roots, gradually cutting off the flow of nutrients and water to the vine. Nymphs also form protective galls on the undersides of grapevine leaves of some Vitis species and overwinter under the bark or on the vine roots; these leaf galls are typically only found on leaves of American vines. Phylloxera attacks the roots of grape vines and eventually kills the plant. It can take up to ten years from the time a plant is infested with phylloxera until the vine finally dies. During that period of decline, the quality

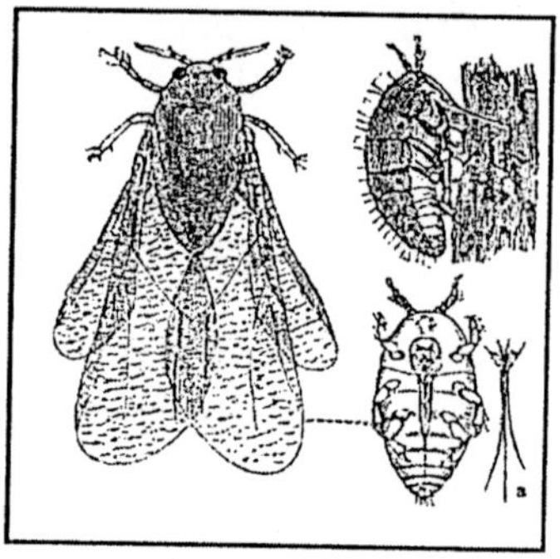

of the grapes produced by the vine remains high though the quantity of fruit produced diminishes each year.

Lifr Cylce

The phylloxera louse has a complex life-cycle of up to eighteen stages that is divisible into four principal forms: sexual form, leaf form, root form, and winged form. The sexual form begins with male and female eggs laid on the underside of young grape leaves. The male and female at this stage lack a digestive system, and once hatched, they mate and then die. Before the female dies, she lays one winter egg in the bark of the vine's trunk. This egg develops into the leaf form. This nymph, the fundatrix (stem mother), climbs onto a leaf and lays eggs (via parthenogenesis) into a leaf gall she creates by injecting saliva into the leaf. The nymphs that hatch from these eggs may move to other leaves, or move to the roots where they begin new infections in the root form. In this form they perforate the root to find nourishment, infecting the root with a poisonous secretion that prevents it from healing. It is this poison which eventually kills the vine.

This nymph reproduces four to seven more generations of eggs (which are also capable of parthenogenetic reproduction) each summer. These offspring spread out to other roots of the vine, or to the roots of other vines through cracks in the soil. The generation of nymphs that hatch in the autumn hibernate in the roots and emerge the following spring when the sap begins to rise. In humid areas, the nymphs develop into the winged form (otherwise, they perform the same role without wings). These nymphs start the cycle over again by either remaining on the vine to lay male and female eggs on the bottom side of young grape leaves, or flying to an uninfected vine to do the same. Many attempts have been made to interrupt this life cycle to eradicate phylloxera, but the louse has proven to be extremely adaptable, as no one stage of the life cycle is solely dependent upon another for the propagation of the species.

Habitat

Claimed to be free of phylloxera attack in its high altitude, rocky habitat. Used to create a white varietal wine reportedly possessed of a mildly aromatic fragrance and, in some cases spritzy, delicately crisp flavour.

Control

Cultural: The root form is effectively managed by using resistant or tolerant rootstocks. However, once an infestation occurs, eradication of phylloxera from a vineyard is unlikely, but steps can be taken to keep the infestation at a tolerable level.

Monitoring: Vineyards should be scouted for the foliar form on infested leaves after shoot length has reached five inches. Small galls will be evident on the underside of the terminal leaves.

Chemical : Control of the foliar form of phylloxera may be achieved by applying insecticides during the bloom stage. Early season control of this pest is critical. Several insecticides provide good control of grape phylloxera, and can be found in the Midwest Small Fruit Pest Management Handbook.

Sawfly

Kingdom: Animalia, **Phylum**: Arthropoda, **Class**: Insecta, **Order**: Hymenoptera, **Suborder**: Symphyta.

Sawfly is the common name for insects belonging to suborder Symphyta of the order Hymenoptera. Sawflies are distinguishable from most other Hymenoptera by the broad connection between the abdomen and the thorax, and by their caterpillar-like larvae. The common name comes from the saw-like appearance of the ovipositor, which the females use to cut into the plants where they lay their eggs. Large populations of certain sawfly species can cause substantial economic damage to forests and cultivated plants.

Life Cycle

Life cycles vary by species, but generally they overwinter as a pre-pupa in a cocoon in the ground or other protected place, pupating in the spring. In early summer, adults lay eggs in or on plant tissue. Larvae develop through several stages (up to 6 instars) before pupating, producing one generation per year. Some species have several generations annually.

Habitat, Food Source(s), Damage

Mouthparts are for chewing. Most sawflies are somewhat host-specific. The elm sawfly feeds on elm and willow. Conifer sawflies (such as the loblolly pine sawfly, *Neodiprion taedae linearis* Ross) can occasionally cause serious injury to pines and other evergreens by feeding on foliage, tunneling into buds or boring into shoots. Larvae of the blackheaded ash sawfly [*Tethida cordigera* (Beauvois)], and brownheaded ash sawfly [Tomostethus multicinctus (Rohwer)], feed on red and white ash. Common sawfly larvae feed together (gregariously) and may hold their abdomens coiled over the edge of the leaf. On broadleaved tree leaves, larvae usually feed on the underside of the leaf. On oaks, *Caliroa* spp. larvae skeletonize leaves, while on pecan sawfly larvae produce round holes of various sizes in leaf tissue. Some species are leaf rollers, web formers, leaf skeletonizers, leaf miners or cause plant galls. Adults can be found on flowers.

Management

If it is determined that control measures are needed, use the following as a guide for selecting the best method:

Mechanical Control - Examples here include methods such as hand picking larvae from plants, physically dislodging them by using forceful water sprays, or other means of nonchemical control. Population size and distribution will determine the effectiveness or suitability of the chosen method.

Biorational Insecticides - Insecticidal soap-best for low populations of young larvae. A note here - *Bacillus thuringiensis* (*Bt*) **will not**control sawflies. Strains of this biological insecticide are effective against various caterpillar pests (larvae of butterflies and moths), but will not control sawfly larvae.

If it is determined that chemical control measures are needed, they should be directed toward the young larvae. Young larvae are much more susceptible to chemical applications than larger, more mature larvae. This is especially true when using biorational products such as insecticidal soap. If larvae are nearly full grown, control measures should not be attempted as chemicals will not be as effective, and most of the damage that the plant will sustain has already been done. Finally, because sawflies often feed in groups, chemical applications should be directed only to the areas they are feeding on; entire tree sprays are unnecessary unless populations are wide-spread throughout the plant. The best control for sawflies or any other pests is preventive measures related to plant health. Correct plant selection, proper site selection when planting, and then continued recommended cultural care will ensure that plants are in excellent health. The better condition a plant is in, the more damage it can tolerate without affecting its health status.

Scale Insect

Kingdom: Animalia **Phylum**: Arthropoda, **Class**: Insecta, **Order**: Homoptera, **Suborder**: Sternorrhyncha, **Superfamily**: Coccoidea, Handlirsch, 1903.

The scale insects are small insects of the order Hemiptera, generally classifi-ed as the superfamily Coc-coidea. There are about 8000 species of scale insects. Among the best-known American scale insects is the cottony-cushion, or fluted, scale, so called because of the grooved, fluffy, white eggs it lays, and the San Jose scale, so called because it was introduced into the United States at San Jose, California. The cottony-cushion scale was introduced from Australia in 1868 and caused great damage to California citrus groves until a ladybird beetle was introduced to hold it in check. The San Jose scale is a round insect that destroys apple, pear, and other fruit trees; it secretes a hard, gray coating that is extremely resistant to sprays. The scales most commonly encountered by the home gardener are the soft scales. The brown soft scale parasitizes many house, garden, and greenhouse plants; the hemispherical scale is a common greenhouse pest. Mealybugs often blanket the stems of garden plants in a manner similar to that of aphids and are characterized by their secretions of white, moldlike wax.

Life Cycle

Most scale insects have separate sexes, and reproduction is internal. The female often lays fertilized eggs within her protective wax coating, where they remain until hatching. After hatching, they undergo a series of molts that will eventually lead to the adult scale. In some groups there is hermaphroditism (one individual with both sexes) or parthenogenesis (the development of unfertilized eggs).

Crawlers

The first instar, which emerges from the egg, is called a crawler, because it moves around the plant where it was born until it finds a suitable new growth into which it attaches its mouth parts and begins feeding. From this moment on, the larvae do not move. The fact that the first instar is the only one to move is reflected in its name. From this point on, males and females follow different pathways.

Female Development

After the crawler stage, females are not likely to move again. In the first molt, some scale females lose their legs and antennae, while in others they are greatly reduced. While the latter can move around if necessary, this rarely happens. Both types of females molt a second time before reaching maturity and, as all hemimetabolous insects, do not pupate. The adult female retains many external characteristics of the larva, such as the lack of wings (a phenomenon called neoteny), but has functional reproductive organs.

Male Development

Males undergo two additional molts and emerge as adults with some special characteristics. They have only one pair of wings, which is rare in insects not belonging to the order Diptera (mosquitos and flies). In addition, they do not have mouth parts and do not feed. For this reason, they have very short lives (a day or two). However, they are mobile and are the ones that find the immobile females to mate.

Abundance and Seasonality

Scale insects are parasites that are present year round, but they are more abundant in the warmer months. They may spend the winter in any stage, but the more resistant ones are the eggs and the mated females (which have not yet laid their eggs). When the temperature rises in the spring and plants start new growth, the eggs begin hatching and the crawlers emerge in large numbers.

Many species of scales such as the Putnam, tesserata scale, cyanophyllum scale, mining scale and lesser snow scale are highly parasitized by tiny wasp parasites. Pin-sized holes in the wax are evidence of parasitism. When scouting the nursery, be sure to take into account the amount of parasitism before applying an insecticide. Scale crawlers are also preyed upon by numerous beneficial insects.

Habitat

Like other homopterans, scale insects have an incomplete metamorphosis with three stages: egg, nymph, and adult. The first nymphal stage is known as a "crawler," because it has legs and actively moves about. After the next molt, however, the legs are lost in most species, and the scale insect becomes sessile, secreting a scale like, waxy covering for protection.

Some species of scale insects are economically important as agricultural pests, because of the severe damage that they cause to some crop plants. This injury is usually associated with mechanical damage to foliage caused by piercing by the feeding apparatus of the scale insect. Damage is also caused by the withdrawal of large quantities of carbohydrates and other nutrients with the sap. The cottony cushion scale (Icerya purchasi) is an important pest of citrus crops in the southern United States, where it has been introduced from its native Australia.

This species is much less of a pest than it used to be, because it has been relatively well controlled by several predators that were later discovered in their native habitat and subsequently introduced to the United States, namely, the vedalia lady beetle (*Vedalia cardinalis*) and a parasitic fly (*Cryptochetum iceryae*). This case is commonly cited as one of the great successes of non-pesticidal, biological control of a serious insect pest. The California red scale (*Aonidiella aurantii*) is another important agricultural pest of western citrus trees. The San Jose scale (*Quadraspidiotus perniciosus*) was introduced to North America from Asia in the 1880s, and is a serious pest of many species of orchard and ornamental trees and shrubs. Various species of mealybugs are also important pests, for example, the citrus mealybug (*Planococcus citri*), and the greenhouse mealybug (*Pseudococcus longispinus*).

Females of the Indian lac insect (*Laccifer lacca*) of southern and southeastern Asia produce large quantities of a waxy substance that is collected, refined, and used to prepare a varnish and shellac. Males of the Chinese wax scale (*Ericerus pela*) secrete relatively large amounts of a white wax, which can be collected for use in making candles. The Indian wax scale (*Ceroplastes ceriferus*) produces a wax that is collected for use in traditional medicine. The tamarisk manna scale (*Trabutina mannipara*) occurs in the Middle East, where it feeds on tamarisk trees (*Tamarix* spp.).

The females of this insect excrete large quantities of honeydew, which in arid regions can accumulate abundantly on foliage, drying into a sweet, sugar-rich material known as manna. Manna is featured in the Old Testament of the Bible, in which it is portrayed as a miraculous food delivered from the heavens, sustaining the Israelites as they wandered in the wilderness after their exodus from Egypt.

Control

Cultural Control

Be sure plants are free of scales before they are placed in the production area of the nursery. Scales cannot fly, therefore they do not readily infest plants as do most other insects.

Chemical Control

Scales, especially armored scales are very difficult to control when mature. Spray applications should be timed to coincide with the crawler stage which is most susceptible to insecticides. Examine plants for live scales by crushing the wax cover. Dead scales do not fall from plants. Select pesticides that have the least effect upon other non-target organisms. For established infestations, apply a second application in two weeks. Horticultural oils are often effective and relatively safe on beneficial organisms. One or two applications of dormant oils should be applied to suppress established overwintering populations.

COCHINEAL INSECT

Domain: Eukaryota, **Kingdom**: Animalia, **Phylum**: Arthropoda, **Class**: Insecta, **Order**: Homoptera, **Superfamily**: Coccoidea, **Family**: Dactylopiidae, **Genus**: *Dactylopius*, **Species**: *Dactylopius coccus.*

The cochineal (*Dactylopius coccus*) is a scale insect in the suborder Sternorrhyncha, from which the crimson-coloured dye carmine is derived. A primarily sessile parasite native to tropical and subtropical South America and Mexico, this insect lives on cacti from the genus Opuntia, feeding on plant moisture and nutrients. The insect produces carminic acid that deters predation by other insects. Carminic acid, which occurs as 17-24% of the weight of the dry insects, can be extracted from the insect's body and eggs and mixed with aluminum or calcium salts to make carmine dye (also known as cochineal).

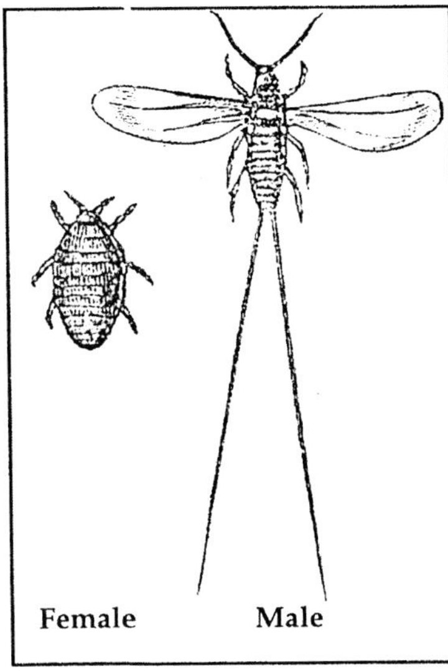

Carmine is today primarily used as a food colouring and for cosmetics. The carmine dye was used in Central America in the 15th century for coloring fabrics and became an important export good during the colonial period. After synthetic pigments and dyes such as alizarin were invented in the late 19th century, natural-dye production gradually diminished. Health fears over artificial food additives, however, have renewed the popularity of cochineal dyes, and the increased demand has made cultivation of the insect profitable again, with Peru being the largest exporter.

There are other species in the genus Dactylopius that can be used to produce cochineal extract, but they are extremely difficult to distinguish from *D. coccus*, even for expert taxonomists, and the latter scientific name (and the vernacular "cochineal insect") is therefore commonly used when one is actually referring to other biological species. The primary biological distinctions between species are minor differences in host plant preferences, in addition to very different geographic distributions. Recent findings that some synthetic red dyes may induce cancer has renewed interest in cochineal production. Cochineal-based dye is again becoming popular as a coloring agent, especially in processed foods.

Life Cycle

The cochineal insect's entire life cycle takes place on the pads of certain species of prickly pear cacti. Although sometimes mistakenly referred to as a beetle, it is in fact a scale insect, a type of bug that is usually quite small and that lives by attaching itself to a host plant and drawing sustenance from it. Like other scale insects, the cochineal produces a protective covering for itself, which appears as a white fluffy material on the pads of the cactus. The females are larger than the males, and live longer than the one week that is typical for males. Cochineal insects produce a chemical called carminic acid, which helps them repel predators, and is the source of the dark purple color used to make cochineal dye.

The traditional method of obtaining the dye is to remove the insects from the cactus pads by hand, and then to dry them in the sun before crushing them into a powder. It's estimated that it takes about 70,000 cochineal insects to produce one pound (about 500 grams) of the cochineal powder. Carmine is a further refinement of the cochineal dye, obtained through a process of boiling the powder with certain other chemical compounds.

Habitat

Deserts and arid areas. Including Xeriscape Landscaping at lower elevations in Arizona. This beetle seems to live at lower elevations which are not subjected to freezing temperatures for long periods of time. Below about 1900 feet.

Range: New Mexico, Arizona, South into Mexico, Northwest to California; also in Montana, Colorado, and Florida.

MEALYBUG

Kingdom: Animalia, **Phylum**: Arthropoda, **Class**: Insecta, **Order**: Hemiptera, **Suborder**: Sternorrhyncha, **Family**: Pseudococcidae, Heymons, 1915.

Mealybugs are insects in the family Pseudococcidae, unarmored scale insects found in moist, warm climates. They are considered pests as they feed on plant juices of greenhouse plants, house plants and subtropical trees.

Life Cycle

Birth

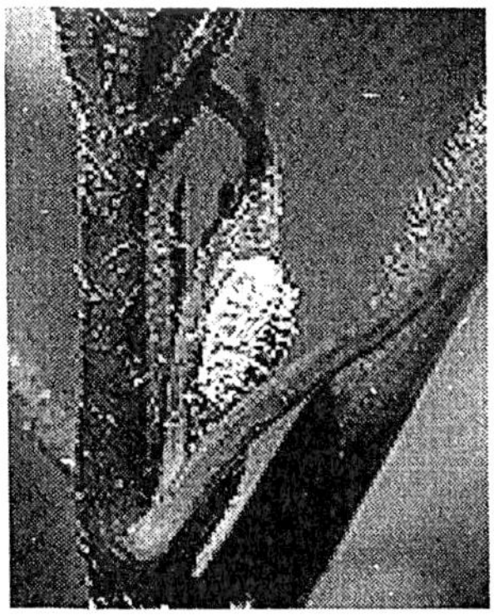

As with most insects, mealy bugs begin life as an egg. The eggs are contained in an egg sac that the female deposits on the underside of a plant leaf where they will be left to mature.

Development

The eggs hatch and the first stage of the mealy bug, known as a crawler or first instar, travels in search of food. They will disperse along the host tree or can be transferred to other trees by hitching a ride on an unsuspecting animal. Second-instar mealy bugs also have the capability to move around, though most prefer to remain sheltered under leaves. It is at this point that the male and female mealy bugs differ in their development. Male mealy bugs will create cocoons and undergo traditional metamorphosis to emerge as winged adults at strategic intervals to ensure successful mating. Females will transition from second instar to third instar before ultimately molting into an adult, skipping metamorphosis.

Reproduction

The mealy bug larvae will mature into adults in about 40 days, when reproduction commences almost immediately. The males will generally only live for two to three days, and the females will remain alive for up to three weeks, allowing them to mature their eggs and lay them before they die. Most of the eggs are laid within 10 days to two weeks of fertilization.

Dormancy

Mealy bugs will produce at least two generations of offspring per summer and spend the winter under the bark of trees including citrus trees, evergreen trees and a host of garden plants. While they do not undergo a state of dormancy, up to 90 percent of eggs are lost during the winter and spring.

Considerations

There is a high rate of mortality for mealy bugs that emerge during the winter and spring months, with only a few first-instar mealy bugs surviving to summer. Maturation tends to occur more slowly with these bugs because they only feed during warm days.

Habitat

The mealy bugs habit citrus orchards, vineyards, sugar cane fields and many other cultivated crops, and are also found on thorn trees.

Control

Mealy bugs are related to scale insects. They cause damage by sucking the juice from their host plants. Like many pests, mealy bugs tend to favor new growth. Over time, their damage causes the leaves to yellow and eventually drop from the plant. They can also cause fruits, vegetables, and flower buds to prematurely drop off. In a bad infestation, their waxy excretions (also known as honeydew) encourages the development of sooty mold fungus. Female mealy bugs hide their eggs in the fluffy white excretions. Eggs hatch in about 10 days, producing crawlers or nymphs. The nymphs relocate to another part of the plant and spend another 4 to 8 weeks developing into the adult form.

How to Get Rid of Mealybugs

Like most pests, the best control for mealy bugs is defensive. Healthy, vigorous plants are less susceptible to infestation than weak, underpotted, and stressed plants. As a general rule, make sure your plants are healthy, and you're less likely to attract these annoying critters in the first place. If you see mealy bugs on your plants, there are several control options:

- **Wash them away**. Mealy bugs can be dislodged with a steady stream of water. Repeat the treatment as necessary. This is best for light infestations.
- **Insecticidal soap**. Insecticidal soaps are available on the market (such as Safer's Insecticidal Soap), or you can make your own by using a dish detergent such as Ivory Liquid. Try to find a product free of perfumes and additives that might harm plants. Mix the soap in a weak concentration with water (starting a 1 teaspoon per gallon and increasing as necessary). Spray on plants.
- **Neem oil**. Neem oil is derived from the neem tree. Use according to label instructions. In addition to its insecticidal properties, neem is also a fungicide and has systemic benefits (meaning the plant absorbs it so it can control insects it doesn't directly contact). According to the Environmental Protection Association, neem is safe for use on vegetables and food plants as well as ornamentals.
- **Kitchen insect spray**. This all-purpose insect spray was developed by the editors of Organic Gardening magazine and came to me by way of Rodale's Organic Gardener's Handbook of Natural Insect and Disease Control. To make a batch, combine 1 garlic bulb, 1 small onion, and 1 teaspoon of cayenne pepper in a food processor or blender and process into a paste. Mix into 1 quart of water and steep for 1 hour. Strain through a cheesecloth and add 1 tablespoon of liquid dish soap. Mix well. The mixture can be stored for up to 1 week in the refrigerator.

A number of stronger insecticides are approved for use against mealybugs, as well as beneficial insects such as beetles

(*Cryptolaemus montrouzieri*) and parasitic wasps (*Leptomastix dactylopii*). However, stronger pesticides can cause a danger to pets and humans, and few indoor gardeners are keen on releasing beetles and wasps in their homes. These biological controls may be better suited to greenhouse cultivation. If an infestation cannot be controlled after two or three weekly applications of insecticide, consider destroying the plant before the mealybugs spread to other plants in your home.

Silverfish

Kingdom: Animalia, **Phylum**: Arthropoda, **Class**: Insecta, **Order**: Thysanura, **Family**: Lepismatidae, **Genus**: Lepisma, **Species**: *Lepisma saccharina*.

Lepisma saccharina, frequently called silverfish, fishmoths, carpet sharks or paramites, are small, wingless insects in the order Thysanura. Its common name derives from the animal's silvery light grey and blue colour, combined with the fish-like appearance of its movements, while the scientific name indicates the silverfish's diet of carbohydrates such as sugar or starches.

Life Cycle

Mating

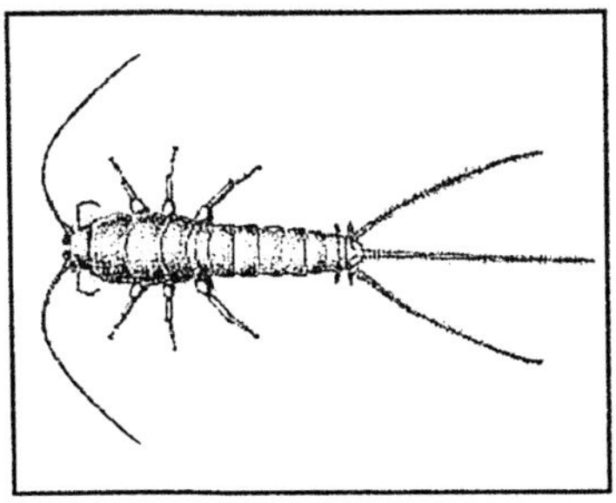

Before mating, silverfish undergo an intricate mating dance of touching and rubbing antennae. The female darts away and then returns to stand beside her mate, who vibrates his tail in a tantalizing mating display. The male deposits his threadlike sperm, which the female picks up and uses to fertilize her eggs.

Eggs

Female silverfish deposit several eggs at a time in a dark place, preferably a nice, damp crack or crevice. The tiny yellowish eggs incubate for a period of about 40 days or so.

Nymphs

Upon hatching, the tiny but fully formed nymphs are white, and will eventually turn gray or silver when they enter adulthood. It takes the nymphs several months to develop into adult silverfish.

Molting

Molting occurs continually throughout the life of a silverfish, though from nymph to adult, a silverfish will need to molt five or six times in order to accommodate its growing body.

Adults

An adult silverfish is about half an inch long, and appears to have silvery scales similar to that of a fish--hence the name. They are fast-moving insects that dislike the light. Throughout their lifespan of several years, they busy themselves filling up on carbohydrates and proteins as found in cardboard, wallpaper, linens and more.

Habitat, Food Source(s)

Silverfish are chewing insects and general feeders but prefer carbohydrates and protein, including flour, dried meat, rolled oats, paper and even glue. They and can survive long periods, sometimes over a year, without food but are sensitive to moisture and require a high humidity (75% to 90%) to survive. They also have a temperature preference between 70 and 80 degrees F. They are fast running and mostly active at night and generally prefer lower levels in homes, but may be found in attics.

Control

Silverfish insects are among the most hated and most feared of common household pests. They are most active at night and sometimes can be seen scampering quickly across the floor at a surprisingly fast speed. They are about a half inch long and have flat bodies. They have long, wispy legs and antennae, and they are usually wider at the head and tape towards their rear. Even though they are so small, their creepy walk and amazing speed makes them very frightening to some people. As pests, they are also filthy and contaminate food storage areas. You can gain silverfish bug control by follow a few different steps to keep them out of your home. Like many pests, the controlling silverfish bugs can be hard if you have a lot of food storage in panties or in basements. You will have to make sure that all areas of your house do not have any loose food, and be sure to vacuum regularly. Just having a lot of crumbs will attract all sorts of pests, like ants and especially silverfish.

If you have dog or cat food, or bird seed, you will need to make sure you have a sealed bin that insects can not chew through. A thick plastic container can help this. Make sure the lid is always sealed. Your kitchen can have a lot of food that will be damaged by silverfish bugs. For silverfish control, you will have to make sure each and every bit of food is covered and sealed. The sugar, flour, oatmeal, and grains should be placed in appropriate containers. A nice glass container will look good in your kitchen and you will be able to keep it sealed so no pests get in. Even a plastic bag or paper bag will not keep the pests out. Many of them can chew through, especially moths. Make sure that everyone keeps all of the containers closed. Cereal boxes are a favorite place for these pests, and many of them lay eggs in them. Place all of your cereal in a plastic, glass, or metal container that they will not be able to crawl into.

When you do find a place in your home that has been invaded by pests, for silverfish control you need to disinfect or throw away whatever has been contaminated. If you have a box of cereal with silverfish in it, you have to throw away the whole thing. You can risk becoming ill or someone in your family getting ill if you do not throw away any infested food.

You should also wipe down your pantries and cabinets with disinfecting wipes that will kill all of the germs that the insects have left behind. Once you have disinfected or thrown away anything that the silverfish have touched, you can create silverfish bug control by using pesticides or bait traps or powders if you still feel as though you have not completely gotten rid of them. You can also use a dehumidifier because they love moisture. Just a few changes can help you with silverfish control.

Snakefly

Kingdom: Animalia, **Phylum**: Arthropoda, **Class**: Insecta, **Infraclass**: Neoptera, (**unranked**): Eumetabola, (**unranked**): Endopterygota, **Order**: Raphidioptera, Handlirsch, 1908

Snakeflies are a group of insects in the order Raphidioptera, consisting of about 210 extant species. Together with the Megaloptera they were formerly placed within the Neuroptera, but now these two are generally regarded as separate orders. Snakeflies are predatory, both as adults and larvae. They can be quite common throughout temperate Europe and Asia, but in North America occur exclusively in the Western United States, namely in the Rocky Mountains and westward, including the southwestern deserts. The adult snakefly has a dark, shiny, flattened head and a prolonged necklike thorax. It can raise its head above the rest of its body, much like a snake preparing to strike. It has long, slender antennae and a brown or dark reddish body that measures 12 to 25 mm (0.5 to 1.0 in) in length. Its four wings are membranous, clear, and heavily veined with a conspicuous spot on the front margins near the tip. When at rest, the wings are held up rooflike over its back.

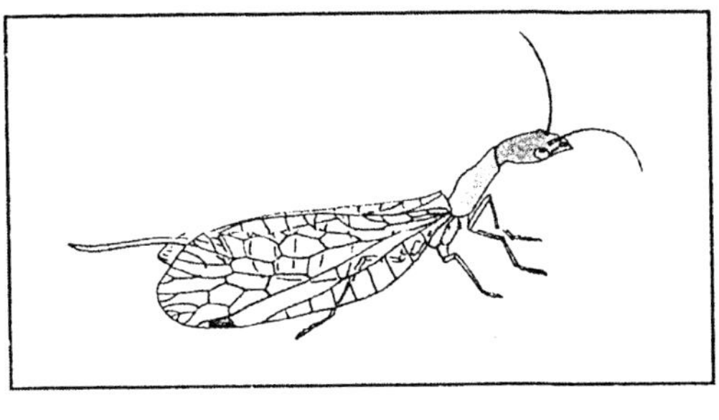

The adult snakefly generally captures only small and weak prey, such as aphids and young caterpillars. The female has an extended, tail-like organ called an ovipositor used to lay clusters of eggs in bark crevices and other hidden areas.

Life Cycle

Snakefly eggs hatch into larvae that live under bark and in leaf litter. Larval feeding habits are not well known, but it likely eats various soft-bodied insects such as wood borer larvae and codling moth pupae. The snakefly larva is flattened and long with a shiny, dark brown to black thorax and head with prominent jaws. It resembles the larvae of certain beetles. Its long abdomen has ten segments and is mostly brown to dark red, commonly with pale spots or rings. The larva undergoes a complete metamorphosis without a cocoon stage and develops through 10 or 11 growth stages, or instars.

Habitat

This species lives in pine, in forests up to elevations of 3,280 feet (1000 metres). The larvae are found underneath the bark of cone-bearing trees, especially pines.

Control

There have been attempts to introduce Snakeflys to Australia and New Zealand for biological control of pests, but they did not become established, possibily because of the lack of a low temperature period to tripper the pupation stage.

Termite

Kingdom: Animalia, **Phylum**: Arthropoda, **Class**: Insecta, **Subclass**: Pterygota, **Infraclass**: Neoptera, **Superorder**: Dictyoptera, **Order**: Isoptera.

The termites are a group of eusocial insects until recently classified at the taxonomic rank of order Isoptera but now accepted as the Epifamily Termitoidae of the cockroach Order Blattaria. Along with ants and some bees and wasps which are all placed in the separate order Hymenoptera, termites divide labour among castes, produce overlapping generations and take care of young collectively. Termites mostly feed on dead plant material, generally in the form of wood, leaf litter, soil, or animal dung, and about 10% of the estimated 4,000 species (about 2,600 taxonomically known) are economically significant as pests that can cause serious structural damage to buildings, crops or plantation forests.

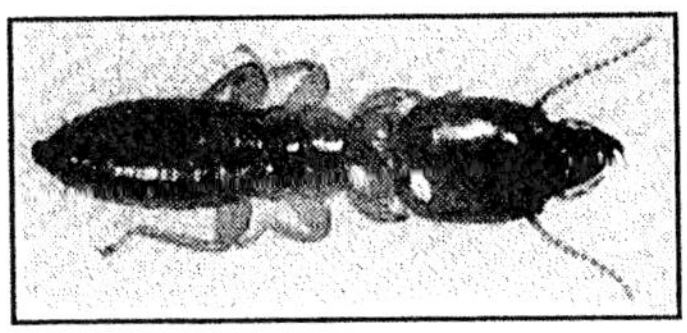

Termites are major detritivores, particularly in the subtropical and tropical regions, and their recycling of wood and other plant matter is of considerable ecological importance. Their role in bioturbation on the Khorat Plateau is under investigation. As eusocial insects, termites live in colonies that, at maturity, number from several hundred to several million individuals. Colonies use a decentralised, self-organised systems of activity guided by swarm intelligence to exploit food sources and environments that could not be available to any the frothy spittle to protect themselves from parasitic and predaceous insects. More than one nymph may be found in a single spittle mass. Spittlebugs have incomplete metamorphosis. Nymphs undergo about five molts and may be orange, yellow, or green. Most species have only one generation per year.

One common species, the meadow spittlebug, is found throughout the United States. It feeds primarily on herbaceous plants, but also occurs on conifers and young woody deciduous plants. Adults are robust and tan, black, or mottled brownish. Nymphs are yellow to green beneath their foaming mass of spittle. The western pine spittlebug is one of several species that are common on conifers and the plants that typically grow under them in the Western United States. Nymphs are dark greenish, brown, or black, sometimes with lighter spots or a pink abdomen. Adults are brownish orange to dark brown and may have an indistinct diagonal white line across the back.

Life Cycle

The life-cycle of the Froghopper is in three stages: egg, nymph, and adult. In this respect it is similar to such insects as grasshoppers and dragonflies. Up to 100 eggs are laid by the adult females in the late summer into an incision made into the tissues of the host plant. The eggs hatch into nymphs the following spring.

The nymph is the sexually immature stage. It resembles the adult in shape but unlike the adult has no wings and only rudimentary legs and eyes. Crucially, it has but a thin outer body layer, or exoskeleton, which unlike the adult's is not covered by a waxy layer. For this reason it needs to protect itself from desiccation. This is achieved by surrounding itself in a frothy mass of soapy bubbles known colloquially as cuckoo-spit. The froth also helps to protect the developing nymph from predators by disguising the nymph. If determined to seek out the insect, the predator will be dissuaded by the unpalatable taste of the froth. (If you do not believe that, try it sometime for yourself!) Enclosed within this froth, the nymph moults several times before emerging as an adult in early summer.

The adult has a hardened exoskeleton enabling it to survive without need for the protection of the cuckoo-spit. They are normally a mottled red-green or brown in colour, although there is variation in hue. The adult has well developed legs which enable it to 'hop' very large distances to avoid predation. They also have a single pair of wings2. After laying its eggs the adult female dies.

Habitat

In nature, the habitats froghoppers are most often found in are woodland edges and grassland. However, they are also a pest known particularly to fruit-growers. They feed on plant sap which they extract from the leaves and stems of plants. This causes minor damage in itself, but the insects carry viruses which can cause serious harm to crops. The eggs are laid on a variety of plants including nettles and grasses as well as the tender young shoots of willow, cherry and apple. In gardens they are frequently encountered on such plants as chrysanthemum, dahlia, fuchsia, lavender, rosemary and rose.

Control

Spittlebugs, or froghoppers as they are sometimes called, are sucking insects that weaken plants, causing stunted or dwarfed growth. Adult spittlebugs lay eggs in field stubble in the fall. Here's how to take control.

Instructions

1. Know what you're looking for. Spittlebugs are oval, ¼ to ½ inch insects that can be brown, black or tan. The nymphs are wingless and form a foamy mass of spittle found in the leaf joints of plants.

2. Look for spittlebugs in late spring or early summer. The foamy spittle is very visible.
3. Wash plants with a strong jet of water at the first sign of the foamy spittle. (Although spittlebugs are not usually a problem to the home gardener, the foam is unsightly.)
4. Spray with a ½ solution of insecticidal soap if the infestation persists.

Springtail

Kingdom: Animalia, **Phylum**: Arthropoda, **Subphylum**: Hexapoda, **Class**: Entognatha (but see text), **Subclass**: Collembola, Lubbock, 1870.

Springtails (Collembola) form the largest of the three lineages of modern hexapods that are no longer considered insects (the other two are the Protura and Diplura). Though the three orders are sometimes grouped together in a class called Entognatha because they have internal mouthparts, they do not appear to be more closely related to one another than they all are to insects, which have external mouthparts. Some DNA sequence studies suggest that Collembola represent a separate evolutionary line from the other Hexapoda, but others disagree; this seems to be caused by widely divergent patterns of molecular evolution among the arthropods. The adjustments of traditional taxonomic rank for springtails reflects the incompatibility of traditional groupings with modern cladistics: when they were included with the insects, they were ranked as an order; as part of the Entognatha, they are ranked as a subclass. If they are considered a basal lineage of Hexapoda, they are elevated to full class status.

Life Cycle

The life of a springtail begins like most other bugs, as a tiny egg. After the egg hatches, there are no larval or pupal stages, instead, they are born as tiny versions of their adult form, thus experiencing what is known as simple metamorphosis.

Maturation

As the springtail grows, it sheds or molts its tough exoskeleton. The main difference between springtails and other insects is that springtails continue to grow and molt after they have reached sexual maturity, up to 50 or more times in their lifetimes.

Ecological Significance

Springtails are ecologically significant because they eat decaying organic matter, contributing to the breakdown and redistribution of dead matter. They tend to prefer fungi and get the majority of their nutrients from the bacteria they ingest while feeding on the decaying matter. Some species eat pollen and green, leafy plant material but do not seem to be a problem. Some are predatory, eating small rodifers and other microscopic organisms.

Environments

The adult springtails must live in areas with high levels of humidity and moisture. Otherwise they will dry up and die quickly. They tend to be found in forest type areas, in rotting logs, compost piles, and other decaying organic habitats. Some species have adapted to being able to survive lower humidity environments but still require a great deal of moisture in order to survive.

Habitat

Springtails live in warm, damp places. They are especially common in leaf litter and soil. Some species are found in caves, the burrows of small mammals, or the nests of ants and termites. They also live in moss, under stones, on the surfaces of ponds and lakes, and along rocky seashores. Others are found high in the trees of tropical rainforests or on the surface of snowbanks. Springtails are sometimes so common in grasslands that there are more than 596,000 individuals per square yard (500,000 per square metre).

Control

Springtail control can be achieved by altering the environment in a way that will reduce the humidity in the area or the moisture content of the soil. In naturally-damp indoor areas this can be accomplished by increased ventilation or artificial air movement from fans. Water leaks in the kitchen, bathroom, or other areas should be repaired since springtails are likely to persist as long as a source of moisture is present. Mildew or mold associated with water leaks should be removed. Wood window sills or other areas with water damage may harbor populations of springtails.

Outdoors, reduced watering and removal of accumulations of leaves or other decaying vegetable matter will help reduce numbers of springtails. If springtails are entering from outside, these efforts should be concentrated around the foundation of the house, doors, windows wells, and other potential entry areas. Springtails in potted plants can be controlled by removal of any algae, fungus, or moss on the soil surface and by reducing watering. Re-potting with clean soil is another option to eliminate insects in the soil. Pressurized aerosols containing permethrin, pyrethrins, or resmethrin can be applied to many types of house plants. Consult the product label for specific plants and directions.

Indoor homeowner-type insecticides labeled for springtail control include certain formulations containing bifenthrin, carbaryl, or diazinon. Many of these are aerosols or ready-to-use formulations. Insecticides for use indoors by commercial applicators include certain formulations containing the above active ingredients, plus some containing chlorpyrifos, cyfluthrin, deltamethrin, or prallethrin.

Outdoor homeowner-type insecticides labeled for springtail control include certain formulations of bifenthrin, carbaryl, chlorpyrifos, deltamethrin, and diazinon. Commercial outdoor insecticides include some formulations of the above, plus some containing fluvalinate, malathion, prallethrin, or pyrethrins.

Stonefly

Description

Nymphs

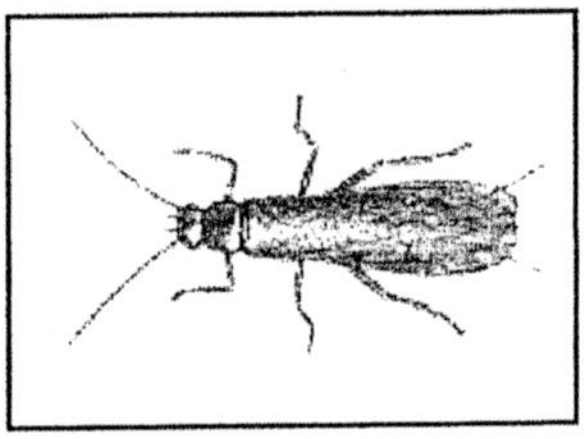

- 5-50 mm in body length
- 2 long antennae
- eyes are widely separated
- gills are usually found on the throat and the base of the legs and abdomen
- 3 pairs of crablike legs with terminal tarsi with 2 tarsal claws
- 2 separate pairs of wing pads
- 2 long tails (cerci)
- with over 460 species known to inhabit North American, colour is varied, however shades of yellow, orange, green, brown and black are the most common. Quite often stoneflies are 2 toned.

Adults

- adult stoneflies look much the same as the nymphs with the following differences:
- wings are folded flat over the back and often extend 10 to 20% past the tip of the abdomen
- 2 short tails (cerci)

Life Cycle

Eggs

Most female stoneflies skim the surface of the stream, dipping their abdomens in the water and releasing their eggs. Others will actually crawl to the bottom of the stream and then release their eggs on submerged objects. Stoneflies are very clumsy fliers and during the egg releasing procedure will cause quite a fluttering and splashing on the water surface which immediately attracts the attention of hungry fish especially steelhead.

Nymph

Eggs hatch into a nymph stage which, depending on the species, can last up to 3 years requiring up to 25 molts to develop completely. Stoneflies do not go through a pupa stage and is therefore considered an incomplete metamorphosis.

Adult

Nymphs will crawl out of the stream onto a stone, tree branch or log and remain long enough to dry and split it's nymphal case. The adults will emerge from the exoskeleton, looking very much like nymphs with wings added and will then fly or climb into the nearby trees. Following the emergence the adults will mate (males attract the females by drumming their abdomens on a tree branch) and then the females will start the lifecycle over again by depositing her eggs back into the stream. Depending on the species, stonefly adults may live for several weeks.

Habitat

Stoneflies live on all continents, except Antarctica. They are also found on most larger islands except Cuba, Fiji, Hawaii, and New Caledonia. There are approximately two thousand species of stoneflies worldwide, with about six hundred in the United States and Canada.

Habitat

Stonefly larvae are found in small streams and medium rivers, but are most common in small streams, including temporary streams and springs. They are usually located in gravel or detritus in sections of fast or moderate flow. In cold climates a few species even fly in midwinter when snow is on the ground. In some locations enormous numbers of adult stoneflies may occur at about the same time each year. Adults can be collected near porch lights to which they are drawn. Stoneflies also commonly swarm along roads, annoying motorists by fouling windshields with their numerous bodies.

Control

Stoneflies cannot be controlled through treatments because they breed in aquatic environments and only become pests when attracted by outdoor lights to buildings. Any emergence of stoneflies, however, should last only a few days. Where these insects are being a problem, exterior light fixtures should be turned off or have yellow "bug lights" bulbs installed. Commercial buildings should use sodium vapour lamps in fixtures rather than mercury vapour lights.

Strepsipteran

Kingdom: Animalia, **Phylum**: Arthropoda, **Class**: Insecta, **Order**: Strepsiptera, Kirby, 1813

The Strepsiptera (known in older literature as twisted-winged parasites) are an order of insects with ten families making up about 600 species. The early stage larvae and the short-lived adult males are free-living but most of their life is spent as endoparasites in other insects such as bees, wasps, leafhoppers, silverfish, and cockroaches.

Male Strpsiptera

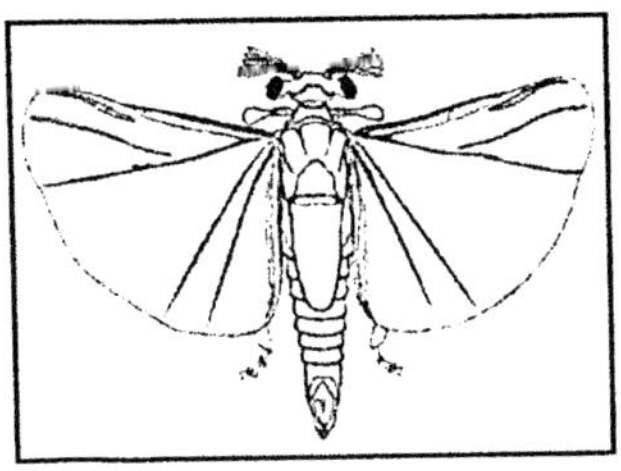

- 1-8 millimetre wingspan
- Large berry-like eyes
- Fan-shaped antennae
- Forewings reduced to knobs (similar to the halteres of flies)
- Large membranous hind wings with reduced venation
- Mouthparts are reduced and non-functional

Female Strpsiptera

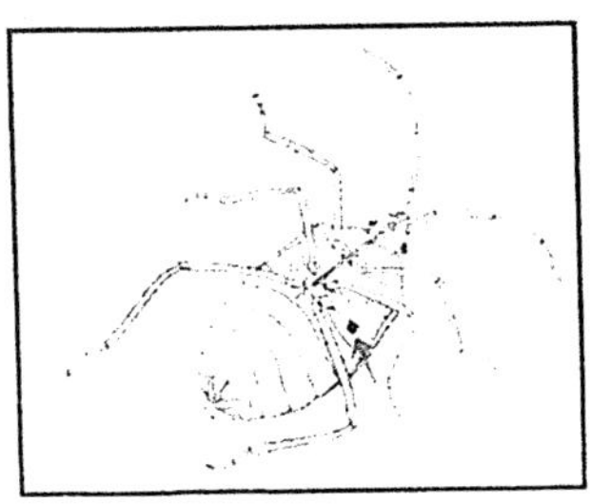

- Larvae-like body
- Wingless
- Legless

- Reduced mouthparts
- Larvae appear grub-like

Life Cycle

Stylopids reproduce sexually and females most likely protrude through the abdomen of the host and release pheromones to attract males. After mating the eggs develop and hatch into first instar larvae within the females body. They then make their way out through her genital opening to find a new host. At this stage the larvae are active, have long thin legs and are able to jump distances up to a few centimetres. After finding a host the larvae burrows in through the cuticle and moults into a legless second instar which completes development within the body of the host. Most stylopids pupate protruding from the host and emerge without killing it. Males will fly off to find a female to mate with while females will emerge and stay on the original host.

Feeding

All species of female and larvae stylopids parasitise a variety of insect orders with most not confined to a single host species. Male stylopids have greatly reduced mouthparts and do not feed.

Habitat

Stylopids are found over much of the country and in Australia they mostly parasitise bugs (Hemiptera) although some other insect orders such as Orthoptera and Blattodea are affected. The free living stages of this order are rarely found. Most male Halictophagus species (above) are parasites of Cicadellidae species (Hemiptera).

Termite

Kingdom: Animalia, **Phylum**: Arthropoda, **Class**: Insecta, **Subclass**: Pterygota, **Infraclass**: Neoptera, **Superorder**: Dictyoptera, **Order**: Isoptera.

The termites are a group of eusocial insects until recently classified at the taxonomic rank of order Isoptera but now accepted as the Epifamily Termitoidae of the cockroach Order Blattaria. Along with ants and some bees and wasps which are all placed in the separate order Hymenoptera, termites divide labour among castes, produce overlapping generations and take care of young collectively. Termites mostly feed on dead plant material, generally in the form of wood, leaf litter, soil, or animal dung, and about 10% of the estimated 4,000 species (about 2,600 taxonomically known) are economically significant as pests that can cause serious structural damage to buildings, crops or plantation forests.

Termites are major detritivores, particularly in the subtropical and tropical regions, and their recycling of wood and other plant matter is of considerable ecological importance. Their role in bioturbation on the Khorat Plateau is under investigation. As eusocial insects, termites live in colonies that, at maturity, number from several hundred to several million individuals. Colonies use a decentralised, self-organised systems of activity guided by swarm intelligence to exploit food sources and environments that could not be available to any single insect acting alone. A typical colony contains nymphs (semi-mature young), workers, soldiers, and reproductive individuals of both genders, sometimes containing several egg-laying queens. Termites are sometimes called "white ants", though they are not closely related to true ants.

Life Cycle

Termites, infamous due to their destructive tendencies, actually have an interesting life cycle that incorporates the caste system. At birth, all termites are categorized as one of the following castes: workers, soldiers or reproductives. Once a termite has been categorized, it starts to develop physical features that aid in completing its caste's duties.

Development

The termite undergoes a common insect metamorphosis, in stages, after birth: from egg, to nymph and finally, adult. This entire process takes about a year to complete. It is important to note that the termite will have reached maturity after five or six months of life and at that time its caste will be decided.

Queen

There is only one queen per colony of termites. It is common for queen termites to lay millions of eggs in a year's time. The queen emits pheromones that prevent any other termites from developing into a queen termite.

Larvae

After the queen lays her eggs, they hatch into larvae. These larvae will molt several times before developing into one caste.

Soldiers

Soldiers are termites that failed to attain sexual maturity. Their primary duty is to defend the colony. To aid in defense, the soldiers form elongated heads and mandibles that act as scissors against their enemies.

Workers

Otherwise known as the destructive termites, the workers are responsible for retrieving food and repairing any damage done to the colony.

Reproductives

These are termites that have reached sexual maturity. These termites either replace the king or queen termite, if they happen to die, or leave the colony and start their own by becoming a king or queen.

Habitat

Termites are found in warmer regions of the world, especially in the tropics and subtropics, or regions that border on the tropics. Most termites prefer to live in warm, humid climates, especially at low altitudes along river valleys and in coastal areas. Other species are found in mountain forests, deserts, and grasslands.

Control

- Eliminate moisture problems in and around the home.
- Keep proper ventilation in both the attic and the crawl space areas.
- Keep standing water away from the foundation.
- Remove potential food sources from around the house (dead stumps, firewood, wood debris and paper.
- Eliminate all wood to ground contact around the exterior perimeter of the structure.
- Rgularly inspect wood decks and wood fences for

activity or damage.

- Keep all much and other ground cover below the sideing around the exterior perimeter of the home.
- Look for warning signs (wings discarded from swarmer termites, mud tunnels, bubbled or cracked paint)
- Seal nail holes and cracks in exposed wood to help prevent easy access by drywood termites.

DAMPWOOD TERMITES

Common Name: Dampwood Termite, **Kingdom**: Animalia, **Phylum**: Arthropoda, **Class**: Insecta, **Order**: Isoptera, **Family**: Hodotermitidae, **Species**: Varies, **Size**: 1/2" to 5/8", **Shape**: Long, narrow, oval, Colour: Brownish, **Legs**: 6, **Wings**: Yes , **Antenna**: Yes.

Dampwood termites are normally larger in size than other termite species. Like Drywood Termites, Dampwood colonies don't have workers. Younger termites called "false workers" do all the work for the colony.

Diet

Dampwood termites like to live and feed in very moist wood.

Habitat

Because they need lots of moisture, Dampwood Termites usually live in damp, dying wood or in houses with leaking plumbing that keeps the wood wet.

Impact

Dampwood termites do not carry disease and don't usually bother buildings because there is not enough water in the wood.

Control

To avoid Dampwood termites, make sure water drains away from your house and keep damp wood away from your home.

DRYWOOD TERMITES

Common Name: Drywood Termite, **Kingdom**: Animalia, **Phylum**: Arthropoda, **Class**: Insecta, **Order**: Isoptera, **Family**: Kalotermitidae, **Species**: Varies, Size: 3/8" to 1", **Shape**: Long, narrow, oval, **Color**: Light brown, **Legs**: 6, **Wings**: Yes,**Antenna**: Yes.

Drywood termites form colonies of up to 2,500 members. Drywood Termite colonies don't have workers. Younger termites, called "false workers", do all the work for the colony.

Diet

Drywood Termites eat wood, wallpaper, plastics and fabric made from plants.

Habitat

Drywood Termite colonies are usually found in dry wood and they do not require moisture or contact with the soil.

Impact

Drywood termites can build nests and dig tunnels in buildings. These tunnels cause major damage because the wooden support beams can become weak and make the building lean or fall down.

Control

- Make sure firewood and scrap wood is stored away from your house.
- Seal all cracks and crevices around the outside of your home.

FORMOSAN TERMITE

Common Name: Formosan Termite, **Kingdom**: Animalia, **Phylum**: Arthropoda, **Class**: Insecta, **Order**: Isoptera, **Family**: Rhinotermitidae, **Species**: Varies, **Size**: 1/2", **Shape**: Long, narrow, oval, **Colour**: Yellowish brown, **Legs**: 6, **Wings**: Yes, **Antenna**: Yes.

Formosan termite colonies can be up to 300 feet long and there can be tens of thousands of termites in a single colony. Formosan termite colonies are divided into three groups: workers, soldiers and reproductives. They are the largest and most destructive kind of termite.

Diet

Formosan Termites eat wood and fabric made from plants. When they eat dead trees, these termites help the environment and make space for new plant life.

Habitat

Formosans live in huge underground colonies, and build mud nests inside the walls of a building. They can also live in boats and buildings.

Impact

Formosan termites eat a lot of wood and can damage your house.

Control

- To avoid Formosan termites, make sure water drains away from your house and keep damp wood away from your home.

SUBTERRANEAN TERMITES

Common Name: Subterranean Termite, **Kingdom**: Animalia, **Phylum**: Arthropoda, **Class**: Insecta, **Order**: Isoptera, **Family**: Rhinotermitidae, **Species**: Varies, **Size**: 1/8" to 1", **Shape**: Long, narrow, oval, Colour: Creamy brown, Legs: 6, **Wings**: Yes, **Antenna**: Yes

Subterranean Termite colonies can have up to 2 million members! Their colonies are divided into three groups: workers, soldiers and reproductives.

Diet

Termites eat wood, wallpaper, plastics and fabric made from plants.

Habitat

Subterranean termites need contact with the soil to survive. They live in underground colonies or in wet areas aboveground. They build tunnels to reach food and every spring, groups of reproductive termites fly off to start new colonies.

Impact

Subterranean termites are the most destructive kind of termite. They can eat a lot of wood and they can cause a lot of expensive damage to a house! They can destroy building foundations, wooden support beams, plastic plumbing pipes, sub-flooring, insulation ... even swimming pool liners and filtration systems! Termites can also injure or destroy living trees and shrubs.

Control

- Don't let water pool around your home's foundation.
- Never leave wood scraps in the yard for them to snack on.

Thrips

Kingdom: Animalia, **Phylum**: Arthropoda, **Class**: Insecta, **Subclass**: Pterygota, **Superorder**: Exopterygota, **Order**: Thysanoptera, Haliday, 1836

Thrips (Order Thysanoptera) are tiny, slender insects with fringed wings (thus the scientific name, from the Greek thysanos (fringe) + pteron (wing)). Other common names for thrips include thunderflies, thunderbugs, storm flies, thunderblights, and corn lice. Thrips species feed on a large variety of sources, both plant and animal, by puncturing them and sucking up the contents. A large number of thrips species are considered pests, because they feed on plants with commercial value. Some species of thrips feed on other insects or mites and are considered beneficial, while some feed on fungal spores or pollen. So far around 5,000 species have been described.

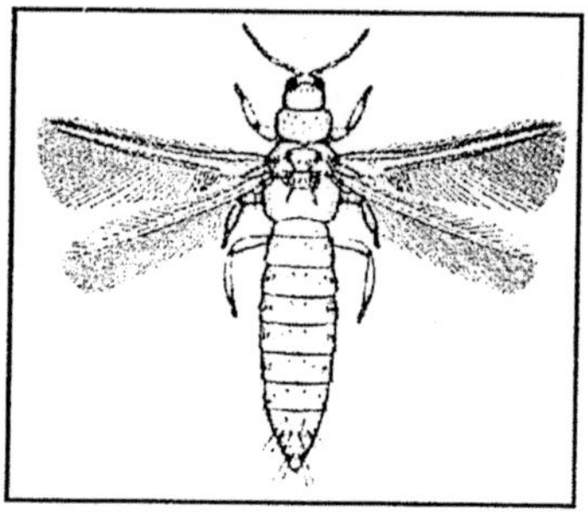

Thrips are generally tiny (1 mm long or less) and are not good flyers, although they can be carried long distances by the wind. In the right conditions, many species can exponentially increase in population size and form large swarms, making them an irritation to humans. Like the words sheep or moose, the word thrips is used for both the singular and plural forms.

So while there may be many thrips there can also be a solitary thrips. The word thrips is from the Greek, meaning wood louse.

Life Cycle

Thrips lay their eggs either on the food plant in slits within the plant tissue or scattered where the female is feeding. On hatching nymphs begin to feed immediately on the plant beside their parents. The nymphs appear similar to the adults but lack wings. Nymphs develop through a few stages before they undergo a pre-pupal and then a pupal stage before becoming fully reproductive adults.

Habitat

Thrips are widely distributed throughout Australia and can be found on a wide variety of plant species, both native and introduced. Many species also occur in leaf litter or under the bark of trees.

Controls

As with any pest, a combination of biological and cultural controls, compatible chemicals and preventative measures gives the best results. The three biological controls described below are compatible and can be used together.

Biological Controls

'Hypoaspis': This soil-dwelling predatory mite feeds on the immature stages of thrips in the soil or growing media. Hypoaspis alone cannot control thrips infestations, but it contributes to the effectiveness of biological control when used with other predators. Hypoaspis can be applied to seedlings before thrips populations become established. It can also be applied to soil under benches in greenhouses.

'Cucumeris': This predatory mite feeds on immature stages of thrips. Release rates for cucumbers and peppers are well established, however, for ornamentals, release rates are still experimental. Crops that produce pollen (e.g., sweet peppers) favour Cucumeris use, because the pollen is a supplemental food for the predator.

'Orius': The tiny pirate bug (Orius spp.) feeds on pollen and all mobile stages of thrips.

- Orius are only effective from March to September because they diapause and do not reproduce if daylength is less than 16 hours.
- Orius should be released once thrips are present in the greenhouse, in areas where thrips numbers are the highest in the greenhouse.
- When Orius is combined with Cucumeris, the thrips predatory mite, relatively low release rates have given excellent control; for example, 1 pirate bug per 1-2 cucumber plants can control an established thrips population in 5 weeks.

Chemical Controls

There are no pesticides for thrips control that integrate well with biological control agents.

- Endosulfan (Thiodan®) has been used at half-rate to knock back a high WFT infestation. It will allow some survival of Cucumeris and Hypoaspis, but will kill all Orius.
- Diatomaceous earth or silica aerogel applied to the floor under the plants will kill thrips as they drop to the ground to "pupate".
- Insecticidal soap or horticultural oil can be used to reduce thrips numbers before biological control agents are applied as they act on contact and do not leave toxic residues.
- Naled (Dibrom®) is effective, but kills all biological control agents and has a residual period of 14 days. It

causes damage to flowers and is best used as a clean-up at the end of the crop.

Other Measures

- Avoid thrips infestations by thoroughly cleaning up the crop at the end of the season. Treat the greenhouse with naled (Dibrom®) after the last pick, before removing crop plants, and again after the greenhouse is empty. When removing plants, do not leave them within or near the greenhouse as surviving thrips will re-enter the
- greenhouse. Wash the greenhouse structure with strong detergent and hot water.
- To starve thrips between crops, maintain a heated greenhouse, empty of plants for 3-5 days, which forces adult thrips to emerge from the soil-borne stages.
- If thrips populations were high in the preceding crop, it may be advisable to apply lime to the ground before covering the greenhouse floor with plastic. Care should be taken to overlap and tape, or glue, the floor covering together.
- Keep ornamental plants out of vegetable greenhouses, and maintain a weed-free border 3 m (10 ft) wide around the outside perimeter of the greenhouse.
- Screen greenhouse vents with thrips-proof screening. This reduces air flow by a factor of 4-5 X, therefore surface area of vents must be increased accordingly.
- Inspect plants brought into the greenhouse for thrips and treat before a biocontrol program is started if many thrips are present.

Whitefly

Kingdom: Animalia, **Phylum**: Arthropoda, **Class**: Insecta, **Order**: Hemiptera, **Suborder**: Sternorrhyncha, **Superfamily**: Aleyrodoidea, **Family**: Aleyrodidae.

The tiny, white, moth-like creatures known as whiteflies, which are familiar to gardeners and flower growers, cotton and vegetable producers, are the adult forms of minute sap-sucking bugs. The eggs are tiny and elongated (about 0.2 mm long), and are laid on young foliage to which they are attached by stalks. These hatch into flattened, oval and scale-like nymphs which bear functional legs in the first instar only. They are almost translucent white with a narrow, waxy fringe around the body, and reach about 0.8 mm in length.

This larval stage, usually found densely spread on the undersides of leaves, is often wrongly assumed to be the egg stage. The nymphs give rise to four-winged adults, which are dusted with a waxy white powder. They are about 1 mm long with a wingspan of not more than 3 mm. They are easily disturbed on the plants and are noticeable when they fly off, literally in clouds, where the infestation is heavy. Bemisia tabaci is probably the most common species found in field crops at lower altitudes, although its taxonomy has become somewhat complicated.

Trialeurodes vaporariarum is likely to be more common in glasshouse and sheltered garden situations, and highveld areas. In the last few years, however, mixed populations of both species have become common in the middle and higher altitudes particularly on field crops such as cotton.

Life Cycle

Greenhouse whitefly, cabbage whitefly and some other species breed all year, especially in warmer parts of the country and in greenhouses. In these situations all life stages may be found all year round. All whitefly have a similar life cycle. The adult females tend to feed on younger, more nutritious leaves. Eggs are laid on end with the tip inserted into the leaf or on their sides. The number of eggs laid per day varies with temperature and suitability of host plant, but a female can lay more than eight per day and between 150 and 300 in a lifetime. Like all insects, whitefly grow by moulting or shedding their skin between each instar or stage.

Crawlers (first instar nymphs) hatch out of eggs. They have three pairs of functional legs and antennae and wander over the leaf until they successfully probe the leaf to obtain sap from the phloem. Once a good feeding site has been found whitefly remain immobile until they become adults. Second and third instar (stage) nymphs are flat, oval-shaped and scale-like. The fourth instar starts off flat and oval like the third instar, but it swells and becomes denser and develops the characteristic appearance of the species. This can include waxy and spine-like processes. The fully grown nymph stops feeding and pupates within the nymphal skin. It is now called a puparium. The adult develops in the puparium. Before hatching the eyes and outline of the body can be seen through the skin.

Adult whitefly force open a T-shaped split in the puparium and emerge. The wings are expanded and grow a covering of wax. After a few hours they can fly and may mate during the next few days. Depending upon temperature, egg laying may start in one to three days. During the first few days they move from the old leaves on which they have hatched to younger leaves on the same or a different plant.

Eggs from unmated females produce only males whereas eggs from mated females may produce females and males. The time of development from egg to adult of each species depends upon temperature and on host plant. Development is faster at higher temperatures and on more favourable hosts.

Greenhouse whitefly. Trialeurodes vaporariorum is the commonest species in New Zealand. It may be found all year round in greenhouse crops and outdoors in the warmer parts of the country. On plants with smooth leaves, eggs are laid in circles around the point of feeding, but on hairy leaves eggs are laid randomly.

After hatching, female greenhouse whitefly take 2 days at 22°C before they can lay eggs (pre-ovipositional period) and on tomatoes at this temperature they can lay 6 eggs per day. At 22°C on tomatoes the time required for each juvenile stage is: egg, 7 days; nymph 1, 4 days; nymph 2, 4 days; nymph 3, 4 days; nymph 4 and pupa, 9 days. Development ceases below 8°C and above 35°C and egg laying stops below 7°C.

Habitat

They are usually found on twigs and the undersides of leaves on a wide variety of tropical plants. They are also widespread on cultivated plants in greenhouses.

Control

Like most pests, the best control for aphids is defensive. Healthy, vigorous plants are less susceptible to infestation than weak, underpotted, and stressed plants. As a general rule, make sure your plants are healthy, and you're less likely to attract these annoying critters in the first place.

If you see whiteflies on your indoor plants, there are several control options:

- **Vaccum them away**. Being careful not to damage the plant, you can use your vacuum cleaner's hose attachment to hoover up adult whiteflies. Make sure that any newly hatched whiteflies can't escape from your vacuum bag!
- **Sticky tape**. The same sticky tape that works for houseflies can be used for whiteflies. Hang it from the infected plant if possible and follow label directions.
- **Insecticidal soap**. Insecticidal soaps are available on the market (such as Safer's Insecticidal Soap), or you can make your own by using a dish detergent such as Ivory Liquid. Try to find a product free of perfumes and additives that might harm plants. Mix the soap in a weak concentration with water (starting a 1 teaspoon per gallon and increasing as necessary). Spray on plants. These will help control the population, but are unlikely to wipe them out.
- **Neem oil**. Neem oil is derived from the neem tree. Use according to label instructions. In addition to its insecticidal properties, neem is also a fungicide and has systemic benefits (meaning the plant absorbs it so it can control insects it doesn't directly contact). According to the Environmental Protection Association, neem is safe for use on vegetables and food plants as well as ornamentals. Like insecticidal soap, neem is useful for controlling whitefly populations, but might not eliminate the problem.
- **Kitchen insect spray**. This all-purpose insect spray was developed by the editors of Organic Gardening magazine and came to me by way of Rodale's Organic Gardener's Handbook of Natural Insect and Disease Control. To make a batch, combine one garlic bulb, one

small onion, and one teaspoon of cayenne pepper in a food processor or blender and process into a paste. Mix into one quart of water and steep for one hour. Strain through a cheesecloth and add one tablespoon of liquid dish soap. Mix well. The mixture can be stored for up to 1 week in the refrigerator.

Zorapteran/ Angel Insects

Kingdom: Animalia, **Phylum**: Arthropoda, **Subphylum**: Hexapoda, **Class**: Insecta, **Subclass**: Pterygota, **Order**: Zoraptera, Silvestri, 1913, **Family**: Zorotypidae, Silvestri, 1913, **Genus**: Zorotypus, Silvestri, 1913

The name Zoraptera, derived from the Greek "zor" meaning pure and "aptera" meaning wingless, was given to the order before winged forms were discovered. The order Zoraptcra is one of the least diverse and most poorly understood of insect lineages. Zoraptera are minute, hemimetabolous insects superficially resembling barklice (Psocoptera) or termites (Isoptera). Individuals are generally less than 3 mm in total body length, exclusive of their antennae, and are gregarious. Overall, zorapterans are quite homogenous in their morphology and tend to be off-white (mostly nymphs) to brown in colour.

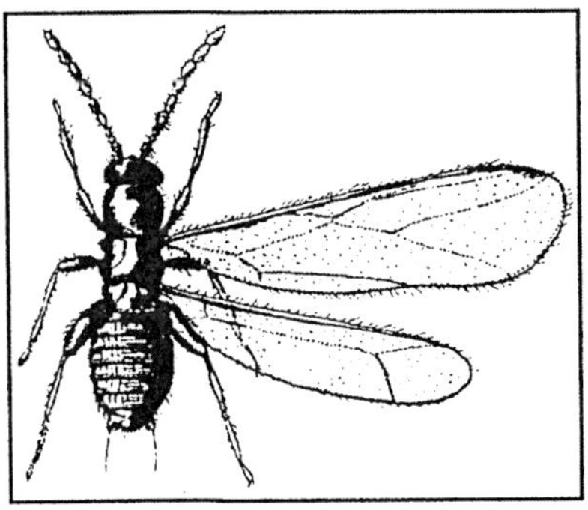

Zoraptera is a small order of exopterygote neopteran insects, for which the only general common name is zorapterans, although "angel insects" occasionally has been used. The affinities of Zoraptera have long been debated, and the order has been regarded as distinct but difficult to place phylogenetically. The major point of contention is whether they are primitive Paraneoptera (and thereby allied with bugs, lice, and their relatives), are related to cockroaches (Blattodea), or perhaps are descendants of some pre-dictyopteran insect stock. Other suggested alliances, based in part on the presence of distinct cerci, are as a sister group of the termites (Isoptera) or the webspinners (Embioptera). Although it has not yet been studied for many species, wing structure allies them most closely with cockroaches and their allies, with recent authors believing them to be primitive Blattoneoptera that separated early from the main lineage. There are no known fossils, other than for one species (*Zorotypus palaeus*) from Dominican amber (Lower Miocene/Upper Eocene).

All Zoraptera have been placed in the single family Zorotypidae. Suggestions that the New World taxa may constitute a separate family, as yet unnamed, need further verification. Until recently, all the 30 described species were allocated to the single genus Zorotypus. In 1993 six additional genera (Brazilozoros, Centrozoros, Floridazoros, Latinozoros, Meridozoros, and Usazoros) were erected, based on wing venation. All of these genera contain one species and are from the New World. All Old World species are placed in Zorotypus. Zorotypus is now a "holding genus" for the many species for which venation is unknown. Species identification often is difficult, with the females of some species almost impossible to place without associated males.

Zoraptera are small, elongate insects, ranging from 0.08 to 0.16 in (2–4 mm) in length. All species are similar in appearance and require dissection and microscopic examination for accurate identification. Adults usually are brown but range from pale brown to almost black and typically are not strongly patterned. The nymphs are pale creamy brown. Two distinct body forms occur in both males and females: wingless individuals (by far the more common and frequently encountered form) and winged individuals, which are generally darker and hold the wings flat over the abdomen. Winged individuals can shed their wings like termites and are recognized by the small wing "stubs" that remain.

The hypognathous head, usually narrowed anteriorly, has chewing mouthparts and a pair of antennae with nine segments. Males of many species have a cephalic gland (fontanelle) on the dorsal surface of the head. Mandibles have several "teeth"; maxillary palpi have five segments, and labial palpi have three segments. The well-defined prothorax is more or less square, and the pterothorax is more developed in winged than in wingless forms. The wings are broad and elongate, with reduced venation, and the hind wing is considerably shorter than the forewing and has less venation. Wingless forms lack both compound and simple eyes (ocelli); winged forms have well-developed compound eyes and three simple eyes. The legs are unremarkable other than for the strongly broadened hind femora, which can bear a row of thickened spines. Their number and arrangement are of taxonomic relevance; the tarsi have two segments with two claws. The oval abdomen is somewhat flattened dorsoventrally, and the short, unsegmented cerci (posterior sense organs) usually have a long apical bristle. Female genital structures are simple, with a well-sclerotized internal spermatheca and a long spermathecal duct. Male terminal structures are more complex, often with distinctive genital details. The pattern of thickened bristles on the posterior apex of the abdomen is also distinctive; the most universal male genital feature is a hooklike projection used to link with the female during copulation.

Life Cycle

Little is known about the biology and life cycle of these insects. They always tend to be gregarious, but there is no evidence of social organisation. Fungal spores and the remains of small mites have been found in the gut contents of dissected specimens.

Habitat

Zoraptera are predominantly subcortical and are thus found under the bark of dead and fallen timber. They appear to be most common in rotting wood and logs with loose bark but with the cambium relatively intact. Their presence may help identify a particular phase in wood decomposition. They sometimes occur with termites. Others are found in ground litter, and Z. *hubbardi* has been reported from piles of decaying sawdust on old milling sites in North America. Winged individuals occasionally are attracted to light or taken in flight interception traps.

Glossary

Abdomen Rear part of insect, the last or hindmost of the three parts of the body of insect, insects the elongated portion of the body of an arthropod, located behind the thorax. It is usually segmented.

Acacia A large shrub or plant with thorns found in warm areas/regions

Acuminate Tapering to a long point.

Acute Quick to respond to even slight stimuli; sharp, extremely serious, severe, or painful

Adaptation Modifications of an organism which enable it to better live in its environment

Adult Developed or fully or matured animal, plant, or other form of life

Adventive Not native and not fully established; locally or temporarily naturalized

Aedeagus Intermittent organ in most insects, formed from a subdivision of the primary phallic lobes.

Aerate to mix with air

Aesthetic Something that is beautiful or artistic rather than useful or practical

Aggregation A group, body, or mass composed of many distinct parts or individuals

Alate Winged; having wings.

Algae A photosynthetic organism of a group that contain

chlorophyll but which are not true plants because they do not have leaves, stems, or roots

Allergic Related to or caused by an allergy

Allomones defensive chemicals used by one species of insect to protect from attacks of another species

Amber Fossilised tree resin

Amblypigids Tailless whipscorpions, Class Arachnida

Ambrosia A kind of fungus that is grown by bark beetles in their tunnels

Ambush Surprise attack

Amphibians Cold-blooded, vertebrate animals that do not have scales; most amphibians lay eggs and go through larval stages during which time they breathe through gills and live in water

Amphibious Able to live both on land and in water.

Amphipods Small crustaceans which have bodies flattened from the sides

Amplify Make larger, greater, or stronger

Anal lobe Posterior lobe of wing.

Anal Pertaining to last abdominal segment which bears the anus.

Annulate Formed in ring-like segments or with ring-like markings.

Anopheline Related to the genus of mosquitoes (*Anopheles*), which transmit malaria to humans

Antenna (pl., Antennae) Pair of segmented appendages located on the head and usually sensory in function - the 'feelers'. Primary organs, used by insects for touching, smelling, and feeling vibrations; for some insects, antennae are used in place of eyes

Antennal Relating to the antenna.

Antivenin Substance that prevents or reduces the effects of a venom

Apiculture Beekeeping; raising and caring for bees

Apophysial pits An exoskeletal pit and evidence of an internal apophysis.

Apophysis An elongate projection from the exoskeleton, either internally or externally.

Appendage Any limb or other organ, such as an antenna, which is attached to the body by arms, legs, tails, etc., attached to bodies

Aptera Primitive wingless insect group.

Apterous Wingless.

Apterygote Any member of the Apterygota - primitively wingless insects (i.e. insects which have never developed wings during their evolutionary history) in modern classifications this includes the Thysanura, but not Collembola, Diplura, and Protura which are no longer considered insects, but are termed Hexapods instead.

Aquatic Related to or living in water

Arachnida Class of arthropods containing spiders, scorpions, mites, and ticks; these arthropods have bodies with two sections, four pairs of legs, and no antennae

Arachniphobia Fear of spiders

Araneae Name for the order that includes spiders; one of several orders in the class Arachnida

Arboreal Living in trees

Arculus. In adult Odonata, an arch-shaped vein connecting radius and median near base of wing, forming an elongate triangle between them.

Arid Excessively dry; having very little rainfall

Arista A long hair that comes out of the third segment of a fly's antennae; some are simply bare hairs while others are feathery

Aristotle Famous Greek philosopher who lived between 384 and 322 B. C.

Aroma Pleasant odour or fragrance

Arthropod From the Latin "Arthropoda," meaning "jointed legs;" includes lobsters spiders, beetles, centipedes, and millipedes; arthropod bodies are made up of segments that are fused together and that are bilaterally symmetrical

Articulate Consisting of segments united by joints.

Assassin One who murders someone

Atrial chamber Cuticular chamber immediately internal to spiracle.

Attractant A substance that attracts insects; pheromones are chemical attractants

Auriculate Ear-like.

Axillary sclerites Sclerites at base of wing.

Bacteria A large group of one-celled microorganisms; some species cause diseases such as pneumonia

Basal half At the base; nearest the body

Batesian mimicry Resemblance of a harmless species to another species that is protected from predators by repellent qualities such as bad tastes; members of the species who mimic the distasteful species are protected from predators because they appear to be distasteful also

Beneficial Of benefit, helpful

Bifurcate Divided into two branches or parts or two-pronged

Bilateral symmetry An ability that promotes forward movement; having matching left and right sides of the body

Biodegradable Something which can be broken down by bacteria or other organisms

Biodiversity Many different species living together in one habitat

Bizarre Extremely odd appearance

Bog Swamp; spongy ground

Book lungs Each book lung has several sheetlike leaves that obtain oxygen from the air: chief breathing organs of spiders located inside the front end of the abdomen;

Bracket fungi A hard, shelf-like fungus that is broadly attached to the trunks of trees

bromeliads Tropical American plants that are usually epiphytes; Spanish moss and pitcher plants are examples

Bubonic Plague A very contagious disease caused by bacteria transmitted by fleas; symptoms include chills, fever, and swelling of the lymph glands

Buccula (pl. **Bucculae)** Flanges on venter of head, on each side of the first segment of the labium in Heteroptera.

Buprestid Member of the family Buprestidae; metallic wood-borers also known as jewel beetles

Burrows To dig such a hole or go under something, as an animal's skin; holes in the ground used as shelter by animals

Callus Applied especially to swollen regions at the front or back of the thorax. A rounded swelling.

Camouflage Using some means of disguise to conceal one's presence

Cannibalistic Eating the flesh of one's own species

Canopy Spreading layer of plant/tree branches at the top of a forest

Capitate With an apical knob-like enlargement.

Carboniferous period Period in the geological time table between 350 million and 260 million years ago, in which large forests grew and became the basis for the production of coal and carbon

Carina A ridge or keel.

Carnivorous Meat-eating animals; eating mainly flesh of other animals.

Carrion Dead flesh

Castes Each caste has certain physical characteristics and carries out specific tasks; divisions of social insects.

Castes The usual three castes are queen, drone (male), and worker. The termites and some of the ants have one or more soldier castes as well.

Cauda The pointed end of the abdomen in aphids.

Caudal Concerning the tail end. Relating to the tail or directed toward the hind part of the body

Cellar An underground room; basement

Cells A cell is closed if it is completely surrounded by veins and opens if it is bounded partly by the wing margin. An area of the wing bounded by a number of veins.

Centipede Arthropods with many segments and with one pair of legs on each segment; the front pair contain poison glands

Cephalothorax The head and thorax are fused together; Part of an arthropod's body that combines the head and thorax.

Cerci (pl. form of cercus) Sensory organs of an arthropod located on the back or hind end of arthropods. The paired appendages, often very long, which spring from the tip of the abdomen in many insects.

Chagas disease Symptoms include high fever, edema, and enlargement of the spleen, liver, and lymph nodes. A tropical American disease caused by pathogens transmitted by species of assassin bugs.

Chelicerae One of two pairs of mouthparts of spiders; these are used to inject venom into the prey

Chromosomes Structures that contain the genes of an individual

Chrysalis In butterflies, the pupal stage between larva and adult when the insect is in a hard case without a cocoon

Ciliate Bearing minute setae.

Ciliation Synonym of chaetotaxy, namely the description of the arrangement of setae.

Circulus (pl., Circuli). An adhesive organ on the abdominal venter of Pseudococcidae.

CITES (Convention on International Trade in Endangered Species), a treaty signed by 21 nations in 1973 to stop illegal trade of plants and animals

Class Insects are in the class Hexapoda in the phylum Arthropoda. A group of related organisms below a phylum and above an order.

Claval commissure Junction of hemelytra along clavus in middle of dorsum, posterior to scutellum in Heteroptera.

Clavus Parallel-sided and sharply pointed anal area of hemelytron.

Claw Sharp curved pincer-like process terminating the segmented thoracic limbs of various arthropods. The sixth segment of insect legs

Cleptoparasite Parasites that steal food that has been stored for the larva of another species

Clypeus Lowest part of the insect face, just above the labrum.

Cobwebs Masses of spider web made by combfooted spiders (family Theridiidae) that are irregular or without a distinct form.

Cochineal dye A red dye made from the dried bodies of female cochineal scale insects.

Cocoon A casing made from silky thread spun by an insect larva to enclose the pupa stage .

Coleoptera Insect order with hard forewings that form a straight line down the back; beetles.

Collophore Ventral tube in Collembola.

Colony Members of the same species that live together as a group.

Compensate Make up for the loss of something.

Compound eye An eye composed of many separate units, each receiving part of the image. Eye made up of an aggregation of separate visual elements, known as ommatidia, each of which corresponds to a single facet of the cornea.

Concave Hollowed or rounded inward.

Conifer Any of a larger order of trees and shrubs that are usually evergreen and that bear cones; examples are pine, fir, hemlock, and yew.

Conspicuous Something that can be easily seen.

Consume Use up; eat.

Contaminate To make impure; pollute

Contiguous Touching - usually applied to eyes (see also Holoptic).

Convex Bulging or rounded outward.

Coriaceous Leather-like in texture.

Corium The main part of the forewing of a heteropteran bug.

Cornea Cuticle covering the compound eye or ocellus.

Cornicle One of the pair of small tubular outgrowths on the hind end of the aphid abdomen.

Cornicle Short, tubelike projection at the back of aphid abdomens used for secreting defensive chemicals

Cosmopolitan Occurring throughout most of the world.

Costa (C) The first longitudinal vein of the wing, usually running along the anterior margin.

Costal Relating to the costa.

Coxa First of eight segments of a spider's leg; first of six leg segments in insects

Coxa The basal segment of the insect leg.

Crepitate In some Acrididae, making a crackling noise by snapping the hind wings in flight.

Crepuscular Active in the twilight.

Cretaceous Geological period in which chalk deposits were formed and flowering plants appeared; this period began 135,000,000 years ago

Crevices Narrow cracks

Cribellate spiders having a cribellum, a structure on the abdomen that can spin a special form of flattened silk

Crochets Hooked spines at tip of the prolegs of lepidopterous larvae.

Crops Plants grown for food

Cross hairs Hairs or fine threads or wires used to define the line of sight when shooting at a target of any type

Crossvein A short vein joining any two neighboring longitudinal veins.

Crustacean Arthropods with hard shells and two pairs of antennae; most crustaceans live in the water, including lobsters, shrimp, and crabs

Cryptic Colouring and or pattern adapted for the purpose of protection from predators or prey by concealment.

Cubital Belonging to the cubitus.

Cubitus (Cu) The sixth longitudinal vein of the wing, immediately behind the media (M).

Cuneus A more or less triangular region of the forewing of certain heteropteran bugs, separated from the corium by a groove or suture.

Cursorial Adapted for running.

Cuticle Outer covering of insect, secreted by the epidermis.

Cuticular Pertaining to the cuticle.

Cylindrical In the shape of a cylinder

Cytotoxin Substance that has a poisonous effect on cells

Deciduous Trees and shrubs that shed leaves each year

Decomposition rot or decay

Defences Ways of guarding against attack

Defuses Make harmless or reduce the effect of

Delicacies Special food; choice morsels

Dens (pl., Dentes) In Collembola, the long proximal segment of the distal arms of the forks of the mannbrium, bearing the mucro apically.

Denticle A conical pointed projection, as a small tooth.

Denticulate Bearing very small tooth-like projections.

Devonian The time before the Carboniferous period. Geological period when plants and animals appeared on land between 405 million and 350 million years ago.

Dicondylar With two condyles or basal attachments to the head capsule.

Diffuses Spreads out widely

Digestive Relating to changing food into material the body can use for nourishment

Digitate Finger-like.

Digitules In Coccoidea, a pair of normally capitate setae at inner base of tarsal claws.

Dimorphic Occurring in two distinct forms.

Diptera Order of insects that have only two wings; includes flies, gnats, and mosquitoes

Disperses scatters about

Diurnal. Active in daytime.

Diversity Being composed of distinct or unlike elements or species; widely different

Docent A guide in a museum or art gallery

Dolbear's Law Formula for finding the temperature using the chirps of a Snowy Tree Cricket; T=50+[(n-40)/4] where "T" is for temperature and "n" is the number of chirps per minute

Domestic Farm animals are domestic animals, as dogs and cats. Living with or cared for by people.

Dorsal. On or concerning the back or top of an animal.

Droppings Bodily waste from animals

Dry forest A forest with trees adapted to long periods of dry weather (or no rainfall); a rain forest would be the opposite

Ecomorphosis A phenomenon where certain life stages are characterized by reduced activity and distinct morphological change.

Ecosystem Habitat composed of many different organisms that live together as one unit

Ectognathous With mouthparts freely exposed beyond head capsule, and visible externally.

Ectoparasite A parasite that lives on the outside of its host.

Elaborate Complicated

Elbowed Bent in the shape of an elbow

Elytra (singluar form is elytron) the hardened front wings of beetles and other insects that cover and protect the hind wings

Elytra The tough, horny forewing of a beetle or an earwig.

Emarginate With a distinct notch or indentation in the margin.

Encephalitis An infection of the brain and nervous system caused by a virus, often transmitted by mosquites

Endemic Restricted to a well defined geographical region.

Endophallus Internal sac or tube of aedeagus.

Ensnare trap

Entognathous With mouthparts sunk into head capsule, and not usually visible externally.

Entomologist Scientist who studies insects; entomologists study insect physiology, behaviour, form and structure, taxonomy, and ecology, as well as pest management and other disciplines that involve insects

Environment Everything in the living and physical world that influences or affects something living

Enzymes Substances produced in living cells that influence a chemical reaction in the body but are not changed themselves; for example, an enzyme helps break down food for digestion

Epidemic Affecting many people at the same time. Rapid spread of a disease.

Epiphytes A plant that grows on another plant; epiphytes get both moisture and nutrients from the air and rain rather than from the soil

Evaporatorium Part of cuticle of thorax, modified for rapid evaporation of scent gland fluid.

Eversible Capable of being turned outward or inside-out.

Evolution Development of a biological group from ancestor to descendant over a long period of time

Excrete Discharge waste from the body; for example, sweat glands excrete sweat

Exoskeleton Collectively the external sclerites of the integument.

Exoskeleton Shell-like body wall found on the outside of an arthropod and moved by internal muscles.

Explanate. Spread out and flattened.

Extinct Something that no longer exists

Exuvia The cast-off outer skin of an insect or other arthropod.

Eyes Organs of sight.

Face Front of head between compound eyes.

Facet The surface of an ommatidium - one of the units making up the compound eye.

Facetted With facets.

Facet The outer part of one of the parts of a compound eye

Facultative Exhibiting an indicated lifestyle under some environmental conditions, but not under others.

Family a group of related plants or animals forming a category ranking above a genus and below an order

Fangs Piercing mouthparts of spiders and other Arachnida that are hollow and used for injecting venom

Feces Animal waste products

Femur (pl., Femora). The 3rd (counting out from the body) and often the largest segment of the insect leg. Also the third leg segment of a spider.

Filament A thread-like structure. Something that is threadlike in appearance

Filiform Thread-like or hair-like, applied especially to antennae.

Flabellate With projecting flaps on one side, applied especially to antennae.

Flagellomere One part of a multi-divided flagellum.

Flagellum The distal (furthest away from the body) part of the antenna, the 3rd primary segment of the antenna.

Flourish To do well or to thrive

Foliaceous Resembling a leaf.

Fontanelle Pore on frontal region of head of termite.

Forage Food for animals; to wander in search of food

Forceps Pinchers at end of body.

Foreleg A front leg.

Forensic related to a court of law

Forewing Either of the anterior wings, borne on the mesothorax. The front wings of an insect having four wings.

Fossil Remains of plants or animals found hardened in layers of rock

Fossorial Adapted for digging.

Frons Upper part of the insect face, between and below the antennae and usually carrying the median ocellus or simple eye.

Furca (pl., Furcula). Forked spring or leaping appendage on fourth abdominal segment of Collembola.

Fusiform. Tapering toward each end.

Gall A lump on any part of a plant that is caused by an injury from insects, bacteria, or fungi

Generation The young produced by the same parents; from one stage in the life cycle to the same stage in the offspring

Genetic Relating to the genes (genes determine the development of characteristics in plants and animals)

Genetics The biological study of the genes and the part they play in heredity

Genitalia The external sex organs. The copulatory organs of insects and other animals. The shape and arrangement of the genitalia are often used to distinguish closely related and otherwise very similar species.

Genus (plural form is genera) A classification of related animals in the taxonomy division ranking above a species and below a family; in scientific name of a species, genus comes first, is spelled with a capital letter, and should be underlined or written in italics

Geometer Caterpillars and moths in the family Geometridae; the caterpillars also are called measuring worms

Gill Breathing organ possessed by many aquatic creatures, including numerous young insects. Insect gills are usually very fine outgrowths from the body and they contain numerous air-tubes, or tracheae. Oxygen passes into the tubes from the water by diffusion.

Girdling Encircling

Glabrous Without setae.

Gland Body organ that forms secretions that can be used by the body or excreted from the body from the body

Glossa The inner most pair of lobes at the tip of the labium or lower lip: usually very small, but long in honey bees and bumble bees, in which the two glossae are used to suck up nectar.

Gonangulum Sclerite attached to base of first valve of ovipositor in female insects, and linking to tergum 1X, and parts of ovipositor derived from segment 1X.

Grooming Taking care of the appearance

Groups of individuals that become irreversibly behaviorally distinct at some point prior to reproductive maturity. One of three or more distinct forms which make up the population among social insects.

Grub Larva of beetle that usually has short legs and is curved in a C-shape

Guano Fertilizer composed of animal waste, especially bat droppings; used as food by some insects

Habitat the place where an animal or plant normally lives

Habitation Home

Haltere One of the club-shaped 'balancers' found on each side of the metathorax among the true flies (Diptera). They are the much-modified hindwings.

Halteres Knob-like structures on the metathoraxes of flies which help to stabilize the insects' flight and to stay on an even course in flight

Haustellum Distal portion of proboscis in adult Diptera.

Head The anterior body region of insects which bears the mouthparts, eyes, and antennae.

Hemelytra One of the basally thickened fore wings of Hemiptera.

Hemimetabolous Having an incomplete metamorphosis, with no pupal stage in the life history.

Hemiptera insect order having piercing-sucking mouth parts and front wings that form an "X" or "V"; true bugs include stinkbugs, bed bugs, squashbugs, and water striders

Herbivorous Plants-eating. Eating only or mainly grass or other plants, or relating to the eating of such plants.

Hexapod Animal with six legs.

Hindleg A back leg.

Hindwing Either of the posterior wings, borne on the metathorax.

Hindwing The back wings of an insect having four wings

Holocyclic Life cycle of aphid species with sexual reproduction on winter host and no migration to a summer host plant.

Holometabolous Having a complete metamorphosis, with larval and pupal stages in the life history.

Homo sapiens human beings

Homoptera Insect order with piercing-sucking mouthparts similar to Hemiptera except the wings rest on the back in the shape of a tent; includes tree and leaf hoppers, scale insects, aphids, and cicadas

Honeydew A sweet part of plant sap that is excreted by aphids, leafhoppers, and other insects

Hormone Internal secretion which controls the activity of some organ or tissue

Host A plant or animal in or on which a parasite plant or animal lives

Host-specific Feeding on only a certain kind of plant or animal

Hover Hang suspended in air

Humidity dampness in the air

Hydrocarbon An organic compound consisting of carbon and hydrogen

Hymenoptera Insect order with two pairs of filmy, membrane-like wings; includes bees, ants, and wasps

Hypermetamorphosis A type of metamorphosis in which the larva changes into different forms of larvae during its development

Hyperparasite A parasite of a parasite

Hypognathous Having a vertical head and face with the mouth-parts at the bottom.

Hypopharyax Median lobe in mouth parts, immediately behind the mouth.

Hypopharyngeal Relating to the hypopharyax.

Imaginal Pertaining to the adult or imago.

Imago The adult insect. The final or adult stage in the metamorphosis of an insect

Incrassate Thickened.

Indicator A tool or instrument used to measure something; some insects are used as "tools" to measure environmental conditions

Infest Trouble or disturb frequently or in large numbers

Insect Arthropod in Class Hexapoda whose body is divided into three sections: the head, thorax, and abdomen; an insect has three pairs of legs and usually one or two pairs of wings

Insectavorous Depending on insects as food.

Insecticide A chemical used to kill insects

Instar The stage in an insect's life history between any two moults. A newly-hatched insect which has not yet moulted is said to be a first-instar nymph or larva. The adult (imago) is the final instar. The stage of an immature insect between moltings; growth stages

Instinctive Inborn in an animal or person as opposed to learned behaviour

Integument The outer layer of the insect. Outer layer that covers an organism or one of its parts; skin and shells are examples of integument

Intercalaries Additional, interpolated or inserted between, as in wing veins.

Intercubital In Plecoptera, crossveins between Cu1 and Cu2 in the forewing.

Intertidal That part of the shore above the low-tide mark

Invertebrates Animals with no backbone or internal skeleton

Iridescence (adjective form is iridescent) The changing of color when moved; reflection of light to form a color different from the pigment of an object

Isopods Members of an order of small crustaceans whose bodies have seven pairs of legs; includes the pillbugs and sowbugs

Isoptera the name for the order that includes termites

Jet propulsion Ejecting air, gas, water, etc., backwards to propel an object forward

Johnston's organ A sense organ located in the second antennal segment of many insects and particularly well developed in male mosquitoes and certain other Diptera.

Jugal Pertaining to the jugum, the basal area of the wing, usually set-off by a jugal fold.

Jurassic Period came just before the Cretaceous period. The geological time period when dinosaurs existed and birds first appeared.

Kingdom One of four divisions of the natural world virus, bacteria, plant, and animal

Knobbed Having round lumps (knobs); butterfly antennae have knobs at their ends

Kunga cake Bread made by natives in Africa from the pressed bodies of midges from the Family Diptera

Labial Concerning the labium.

Labium The mouthparts, or appendage on the sixth head segment, i.e. lower lip-like part of an insect's mouthparts

Labrum The 'upper lip' of the insect mouth-parts: not a true appendage, but a movable sclerite on the front of the head.

Lac A sticky substance left on trees by scale insects; this substance is used in making varnish and shellac

Lacinia The inner branch of the maxilla, the outer one being the galea

Lamellate Having plates, scales, or layers

Larva (pl. larvae). Name given to a young insect which is markedly different from the adult: caterpillars and fly maggots are good examples. Stage of an insect between the egg and pupa stages

Lateral Concerning the sides.

Leaf miner Insect larva that lives between the top and bottom surfaces of a leaf; discolored areas on the surface of the leaf demonstrate the leaf miner's presence

Lepidoptera Insect order that includes butterflies and moths; insects in this order have microscopic scales on each wing

Lichens A plant-like organism that grows in patches on trees, rocks, etc., which is actually a fungus and algae growing together

Ligula Name given to the lobe at the tip of the labium: usually divided into glossae and paraglossae.

Ligulate Strap-shaped.

Linear Elongated with nearly parallel sides.

Locusts Species of grasshoppers which destroy crops as they migrate from place to place in swarms

Lorum (pl., Lora) Plate lateral to the clypeus in Hemiptera.

Lubricates Oils or greases something to make it run smoothly

Luciferin An organic substance that is used by organisms such as fireflies to help produce light without heat

Macrolepidoptera Large moths and butterflies

Macropterous Having long or large wings.

Macroseta A seta conspicuously larger than adjacent setae.

Maggots The larvae of flies which are wormlike in appearance and which live in decaying matter

Maintenance Keeping something in good repair; supporting

Malaria Disease transmitted by *Anopheles* mosquito bites and characterized by chills, fever, and sweating

Mammal Warm-blooded animals that are vertebrates and usually have hair; female mammals produce milk in mammary glands

Mandible The jaw of an insect. It may be sharply toothed and used for biting, as in grasshoppers and wasps, or it may be drawn out to form a slender needle as in mosquitoes. Mandibles are completely absent in most flies and lepidopterans.

Mandibles Jaws; mouthparts in insects used for capturing prey and biting

Mandibulate Having mandibles suited for biting and chewing.

Mange A skin disease which causes scabs, itching, and hair loss

Mangrove A tropical tree or shrub that has prop roots that appear to be extra trunks; these trees and shrubs are important in coastal land building

Manubrium. In Collembola, the large median base of the fucula.

Maturation The process of becoming full-grown

Maxilla (pl., Maxillae) One of the two components of the insect mouth-parts lying just behind the jaws. They assist with the detection and manipulation of food and are often drawn out into tubular structures for sucking up liquids.

Media (M) The fifth longitudinal vein of the wing.

Membranous Of or like a membrane

Meron Posterior part of proximal rim of coxa.

Mesonotum The dorsal sclerite of the 2nd thoracic segment - the mesothorax: usually the largest thoracic sclerite.

Meso-precoxal bridge Precoxal part of the pleuron, anterior to the trochantin in the mesothorax.

Mesothorax The middle segment of the three segments of the thorax of an insect

Mesothorax The second segment of the thorax.

Mesotrochanter Trochanter of middle leg.

Metallic Something that looks like or has the characteristics of metal

Metamorphosis. Name given to the changes that take place during an insect's life as it turns from a young animal to an adult. These changes may be gradual and not too large, as in the grasshopper, and metamorphosis is then said to be partial or incomplete. **Metamorphosis, complete** Process in which an insect progresses through several distinct stages: egg to larva to pupa to adult. **Metamorphosis, incomplete** Process in which an insect egg becomes a nymph, which then grows gradually into an adult.

Metanotum The dorsal sclerite of the metathorax. It is often very small and its sub-divisions are usually obscured.

Metatarsus Sixth of eight segments of a spider's leg between the tibia and the tarsus; insects do not have this leg part

Metathorax The third and last segment of the thorax. The back segment of the three segments of an insect's thorax

Microhabitat The specific environment in which an animal lives

Microlepidoptera Moths those are small in size

Microorganisms An organism that can only be seen through a microscope

Micropterous Small or vestigal wings, being more reduced than brachypterous.

Microscopic Tiny; invisible without the aid of a microscope

Midges Small, two-winged flies

Migrates Moves from one place to another as the seasons change

Millennia (plural form of millennium) Periods of 1000 years

Miller Moth whose wings appear to be covered with dust as the clothes of a person who works in a grain mill would be

Millipede Small, wormlike animal that has two pairs of legs on each of its many body segments; unlike the centipede, a millipede has no poison fangs

Mimic Comething or someone that attempts to resemble someone or something else

Mimicry Close resemblance of one organism to another similar organism or to natural objects among which it lives for protection to the organism

Miniature Small in scale

Minute Tiny. Very small.

Molar, mola Grinding surface of mandible.

Mold Wooly or furry growth caused by fungus

Molecules Very small particles of matter

Molt To shed the exoskeleton, skin, feathers, etc., before a new growth

Moniliform Antennae composed of bead-like segments, each well separated from the next.

Monocondylar With one condyle or basal attachment to the head capsule.

Monoecious Having male and female sex organs in the same individual.

Morsels Small pieces of food

Mouthparts Structures or appendages near the mouth, adapted for use in gathering or eating food.

Mucro The abrupt sharp terminal segment of the furcula in Collembola.

Multiocular Having many openings.

Naiad Young of an aquatic insect (mayfly, dragonfly, damselfly, or stonefly)

Nectar Sweet liquid found in flowers

Nematode Roundworm, usually microscopic, parasite of plants and animals

Neotropics Relating to the Neotropical region which extends south, east, and west from the central plateau of Mexico, and includes Central America and South America

Neurotoxin A poison that acts on the nervous system

Nits Eggs of lice

Nocturnal Active at night

Nocturnal Active at night.

Nodus The kink or notch on the costal margin of the dragonfly wing. The name is also used for the strong, short cross-vein just behind the notch.

Notal Pertaining to the notum.

Notum Dorsal sclerite of thoracic segment.

Nutrient Something nourishing

Nymph An immature insect in incomplete metamorphosis; it differs from the adult in size and in its incompletely developed wings and genitalia

Nymph Name given to the young stages of those insects which undergo a partial metamorphosis. The nymph is usually quite similar to the adult except that its wings are not fully developed. It normally feeds on the same kind of food as the adult.

Obligatory Not optional.

Obtuse Not pointed or acute.

Ocellus (pl., Ocelli) One of the simple eyes of insects, usually occurring in a group of three on the top of the head, although one or more may be absent from many insects.

Ochraceous Pale brownish-yellow in colour.

Odonata Insect order with slender bodies, large eyes, and two pairs of equal-sized, transparent wings; includes dragonflies and damselflies

Ommatidia (plural form of ommatidium) Scientific name for the individual parts or lens in each compound eye

Omnivorous Feeding on both animal and vegetable substances/ food.

Omphalium Metasternal opening of metathoracic scent gland in adult Gerridae.

Ootheca An egg case formed by the secretions of accessory genital glands or oviducts, such as the purse-like structure carried around by cockroaches or the spongy mass in which mantids lay their eggs.

Operculate Having a body process or part that suggests a lid.

Order A category in plant and animal classification that ranks above the family and below the class

Organic Of or obtained from plants or animals

Orifice An opening, as a vent, mouth, or hole through which something may pass.

Ornamental A plant cultivated for decorative purposes rather than as food or for lumber, etc.

Orthoptera Insect order with straight, papery wings, and jumping hind legs; includes crickets, grasshoppers, and locusts

Ostiole A small bodily aperture, orifice, or pore.

Oviparous Producing eggs which are hatched outside the body of the female. Relating to egg production outside body. Birds, fish, reptiles, and insects that reproduce by means of eggs that develop and hatch outside the mother's body.

Ovipositor The tubular or valved egg-laying apparatus of a female insect: concealed in many insects, but extremely large among the bush-crickets and some parasitic hymenopterans. A tubular organ at the end of the abdomen of some female fish or other organisms, especially insects, that is used to deposit eggs. An appendage (or structure) in certain female insects used to deposit eggs

Ovisac A waxen sac into which eggs are laid in Coccoid Hemiptera, and which sometimes encloses all or part of the female.

Pala Tarsus of foreleg of Corixidae, modified into a seta-fringed scoop.

Palp A segmented leg-like structure arising on the maxilla or labium. Palps have a sensory function and play a major role in tasting food.

Papilla A small projecting body part similar to a nipple in form.

Parabollically With two or more non-parallel sides.

Paraglossa The outer most pair of lobes at the outer edges of the tip of the labium: with the central glossae, the paraglossae make up the ligula.

Paramere Copulatory hooks formed from outer subdivision of primary phallic lobes.

Paraproct One of the 2 lobes bordering the sides of the anus.

Parasite Plant or animal living on or in another plant or animal from which it gets its food; the parasite often injures the host animal or plant; plants or animals that are parasites are said to be "parasitic"

Parasitic Of or like a parasite

Paratergite Lateral subdivision of tergum.

Parthenogenesis Reproduction with unfertilized eggs

Parthenogenetic Reproducing by parthenogenesis, that is with egg developing without fertilization.

Pasteurization Process of using heat to kill harmful bacteria in things such as milk

Patella Fourth of eight segments of a spider's leg; in human beings the patella is the kneecap; insects do not have a patella

Pathogen Something capable of producing a disease, especially a living microorganism or virus

Pedipalps One of two pairs of mouthparts in spiders; the base of pedipalps can be used to crush food; in males, the tips are swollen and modified to transmitting sperm during reproduction

Penis (pl. Penes) Intromittent organ formed from primary phallic lobes without subdivision. Male sex organ. The external male organ of copulation, used to transfer semen to the female. In most mammals, it is also used to expel urine from the body.

Pentagonal. Having five sides and five angles.

Penultimate The next to the last member of a series.

Perception An understanding through observation

Pesticide Chemical used to kill pests

Petiole The slender stem that supports the flat part (blade) of the leaf

Phenomenal Remarkable

Pheromone Any chemical substance emitted by members of a species that stimulates other members of the species to respond in a particular way

Philatelist A stamp collector

Photosynthesis The process by which organisms convert sunlight to energy

Phragma (pl. Phragmata) Internal plate providing attachment for longitudinal muscles in meso and metathorax.

Phylogenetic Having natural evolutionary relationships.

Phylum (plural form is phyla) Division in the animal taxonomy that ranks below kingdom and above class; the phylum Arthropoda includes spiders, insects, and related animals

Pigments Natural colouring for tissue. A substance that is added to give something such as paint or ink its color. Pigments are often available in the form of dry powders to be added to liquids. Substances that give any object color, such as the scales on wings of butterflies and moths. Colouring substance.

Pilose Densely clothed with setae.

Pistil The female reproductive plant part. The female part of a flower where the eggs are produced. a carpel or group of fused carpels forming the female reproductive part of a flower and including the ovary, style, and stigma.

Plague The epidemic disease. Contagious disease that often is fatal; one form, the bubonic plague, referred to as "Black Death," is caused by bacteria that infect rats and is transmitted by fleas

Pleura Pleural of pleuron. The lateral region of a thoracic segment. Membrane around lungs. A thin transparent membrane that lines the chest wall and doubles back to cover the lungs, thereby forming a continuous sac enclosing the narrow pleural cavity. The inner faces of the cavity are lubricated by fluid to ease breathing movements.

Pleurodema External sulcus of pleural apophysis.

Plicate Having the surface thrown up into or marked with parallel ridges, being folded or pleated.

Pollen The fertilizing powder. A yellowish powder produced by the male parts of flowers and used for fertilization. A powdery substance produced by flowering plants that contains male reproductive cells. It is carried by wind and insects to other plants, which it fertilizes.

Pollination transfer of pollen from one flower to another. To transfer pollen grains from the male structure of a plant anther to the female structure of a plant stigma and fertilize it.

Polyembryony Multiple embryo production. the production of more than one embryo from a single egg.

Polymorphic A species having several forms independent of the variations of sex.

Polyphagous Feeding on a variety of plants and or animals.

Porrect Extending horizontally forward: applied especially to antennae.

Postantennal angle In Anoplura, angular part of head behind the antenna.

Potential Possible, latent, probable, likely.

Predaceous (also predacious)—Hunting and killing other animals. Attacking other animals. Animals that hunt, kill, and eat other animals.

Predator Killer, marauder. Carnivorous animal or destructive organism. A carnivorous animal that hunts, kills, and eats other animals in order to survive, or any other organism that behaves in a similar manner.

Prehensile Adapted for seizing or grasping especially by wrapping around. Able to grasp something. Able to take hold of things, especially by wrapping around them.

Prehistoric Ancient, primitive, early, antediluvian, from periods before written history

Prey An animal or animals caught, killed, and eaten by another animal as food.

Primary Chief, main, prime, major.

Primary phallic lobes Pair of ectodermal outgrowths that give rise to all or part of intromittent organ in insects.

Primitive Prehistoric, Ancient, Primordial. Archaic from long ago; a very old form of something

Proboscis Nose, Snout, Feeler, Antenna, An insect's tubelike mouthpart which is used for piercing and sucking

Proboscis. Name given to various kinds of sucking mouths in which some of the mouth-parts are drawn out to form tubes.

Prognathous. Having a more or less horizontal head, with the mouth-parts at the front.

Proleg. Hollow, paired, non-segmented outgrowth of abdominal segments of holometabolous larvae, usually used in locomotion.

Prologs legs that appear on the abdominal section of insect larvae but not on the adult

Pronotum the top of the first segment of the thorax

Pronotum. The dorsal sclerite of the 1st thoracic segment.

Prosternal Pertains to the sternum of the prothoracic segment.

Protest single-celled organisms that are neither plants nor animals but that have characteristics of both

Prothorax the first segment of the thorax; the prothorax is connected to the head

Prothorax The 1st or anterior thoracic segment.

Protozoan A single-celled organism that can move and feeds on organic compounds of nitrogen and carbon, e.g. an amoeba. Kingdom: Protoctista. A protist with animal-like characteristics

Pseudocellus Thin walled oval structure on the head of Collembola.

Pseudoscorpions A minute eight-legged organism arthropod. Tiny arachnids that have crab-like pinchers but no tail stinger and that feed on insects, mites, and other small invertebrate animals.

Pterostigma A small coloured area near the wing-tip of dragonflies, bees, and various other clear-winged insects: also called the stigma.

Pulp Soft tissue, Fleshy tissue, Flesh. The part that is left when most of the liquid is pressed out of fruit, vegetables, etc.

Pulsing regular beat of blood flow.

Pulvilli (singular Pulvillus) Bladder-like appendages arising ventrally between claws.

Punctate Covered with tiny pits or depressions, like the elytra of many beetles and the thoraxes of many hymenopterans.

Pupa (pl., Pupae) The 3rd stage in the life history of butterflies and other insects undergoing a complete metamorphosis during which the larval body is rebuilt into that of the adult insect a non-feeding and usually inactive stage.

Pupa Cocoon, Chrysalis. Developing insect inside cocoon.

Pupal Related to pupa. An insect at the stage between a larva and an adult in complete metamorphosis, during which the insect is in a cocoon or case, stops feeding, and undergoes internal changes

Puparium dried skin of a fly maggot that surrounds the pupa in much the same way as a cocoon surrounds a moth pupa

Pupate Become pupa. The develop from a larva into a pupa. Enter the pupa stage of metamorphosis.

Pygidium. A caudal structure or the terminal body region of various invertebrates.

Pygmy Of small breed. Belonging to a small breed. Any very small animal or thing

Pyriform. Having the form of a pear.

Quadrangle. Four-sided discoidal cell in Odonata wing, positioned between the anterior median and posterior cubitus, and immediately distal to posterior segment of arculus.

Quadrate. In the form of a square.

Radial. Arranged or having parts arranged like rays.

Radius (R) Third longitudinal vein of wing.

Rafts a mass of some type (such as insect eggs) that floats

Rangeland A land on which cattle, sheep, etc., may graze

Raptorial Using other animals as prey. Adapted for catching prey.

Raptorial Adapted for seizing and grasping prey, like the -front legs of a mantis.

Recluse somebody living apart from others; a solitary person who avoids other people.

Rectal. Relating to, affecting, or being near the rectum.

Rectum Lower part of intestine, the end or lowest part of the large intestine.

Rectum In insects, the posterior expanded part of the hindgut, typically pear shaped.

Regenerated Form again, recover from decline, grew back; formed new tissue

Regurgitates Bring food from stomach, flow out, vomits.

Repellent Disgusting, revolting, hideous, repugnant, making somebody feel intense dislike, disgust, or revulsion, causing disgust.

Repertoire Range, collection, catalog, gamut, a supply of skills or abilities

Replete amply or fully or abundantly provided or filled; worker ant that stores liquid food in its abdomen; ant repletes remain permanently in the nest while other workers gather and transfer food to them (example: honeypot ants).

Repose To lay at rest. Relax. Take it easy. Calm.

Reproduction The act or process of reproducing of offspring or young

Resonate Reverberate, vibrate, continue to sound; vibrating sound

Reticulate Covered with a network pattern. Have network structure. Transitive and intransitive verb to form a network, or be formed into a network. Mark something with lines.

Rhinaria (sin. rhinarium) Sense organs on the antennae of in aphids.

Rodent small gnawing mammal, mammals that have two teeth in each jaw that are used for gnawing wood or similar material; rats, mice, squirrels, and beavers are rodents.

Roost Place where animals sit to rest, settle, sleep; the act of resting at night

Rostral Of or relating to a rostrum.

Rostrum A beak or snout, applied especially to the piercing mouth-parts of bugs and the elongated snouts of weevils.

Rotary Capable of rotating, or turning entirely around.

Saliva Spit, liquid produced by the glands in the mouth

Salivary gland Any gland in mammals that produces and secretes saliva into the mouth, glands in the mouth that secrete saliva

Savanna A flat grassland, sometimes with scattered trees, in a tropical or subtropical region, a grassland with scattered trees

Scale insect A plant-sucking insect that covers itself with a waxy secretion resembling scales. Superfamily: Coccoidea. Small insects in the order Homoptera that feed on and often destroy plants by piercing them and sucking the sap; females cover their bodies and their eggs with a secretion that forms scales

Scale A flat unicellular outgrowth of the integument.

Scales The hollow hairs covering the wings of moths and butterflies.

Scarab type of beetles; these beetles were sacred to the ancient Egyptians

Scavenger An organism that feeds habitually on refuse or carrion.

Scavengers an animal that lives on decaying organic matter

Sclerite A sclerotized plate in the exoskeleton or integument.

Sclerotized Hardened especially by the formation of sclerotin: an insoluble tanned protein permeating and stiffening the chitin of the cuticle of arthropods.

Scutellum The 3rd of the major divisions of the dorsal surface of a thoracic segment: usually obvious only in the mesothorax and very large in some bugs.

Secretions A substance that is produced by some part of an animal or plant

Secretive Not open; living in secrecy

Sedentary Inactive, Sitting, Doing or requiring much sitting

Segment One of the rings or divisions of the body, or one of the sections of a jointed limb.

Senses The five main senses are sight, hearing, touch, taste, and smell, ways in which an organism knows what happens around it. Feeling derived from senses.

Sensillum (plural form is sensilla) Insect sense organ

Sensoria. Circular sensory organs on antennae or legs.

Serrate Toothed like a saw.

Sessile Attached to one place and unable to move, like many female scale insects.

Seta (pl. Setae) A sclerotized hair-like structure of the insect cuticle, arising from a single cell, and surrounded at base by thin cuticular ring.

Setae (plural form of seta) Hairs that cover flies; bristlelike parts or organs found on animals and plants

Setose With setae.

Sheath The covering of a plant or animal

Shellac Varnish, purified resin, A liquid used to give a shiny, smooth appearance to wood and other surfaces. Yellowish orange flakes of a resin lac secreted by a tropical insect.

Simple eyes A single lens that tells the difference between light and dark

Siphoning Draw liquid through tube. Using a tube-shaped organ for drawing in and expelling liquid

Skipper butterfly that moves (or skips) rapidly from flower to flower

Spatulate Shaped like a spatula or spoon.

Species Class, type, kind, sort, genus. A group of animals that can reproduce fertile offspring; species ranks below genus in classification divisions; a species name is spelled with a lower case letter and is underlined or in italics

Spermatophore. A packet of sperm.

Spiderling young spider

Spiders Eightlegged animalthat spines webs. Arthropods with eight legs, two body sections, no antennae, and no wings in the Order Araneae; organs in their abdomens produce a silky thread for spinning webs which are used to catch insects for food, making egg cases, etc.

Spine A multicellular, thorn-like process or outgrowth of the integument not separated from it by a joint.

Spines Sharp, projecting part on an animal.

Spiniform In the form of a spine.

Spinnerets Organs used by spiders for spinning the silk used to make webs.

Spinule. A minute spine.

Spinulose Having minute spines.

Spiracle One of the breathing pores - openings of the tracheal system - through which diffusion of gases takes place. They usually occur on the third thoracic segment and all the abdominal.

Spiracles Small openings through which insects breathe.

Spirals The parts of a spider's web that widens or gradually goes out from the center of the web; the spokes of the web go across the spirals.

Spores "seeds" of flowerless plants such as fungi.

Stamen The male part of a flower that contains the pollen.

Stemmata The simple eyes in holometabolous larvae. Also called lateral ocellus.

Sterile incapable of producing offspring

Sternum Ventral sclerite of a segment.

Steroids Organic/chemical compounds that include sex hormones and cholesterol

Stigma (pl. Stigmata) Small, thickened, and coloured area near the wing-tip of dragonflies, bees, and various other clear-winged insects: also called the pterostigma.

Stresses Strain felt by somebody pressures or forces

Stria Groove running across or along the body: applied especially to the grooves on beetle elytra.

Striate Marked with grooves running across or along the body.

Stridulate To make a sound by rubbing two ridged or roughened surfaces together.

Stridulation To make a chirping or grating sound by rubbing parts of the body together, as male crickets and grasshoppers do; process of making a shrill, grating noise by rubbing together parts of the body

Stridulation The production of sounds by rubbing two parts of the body together: best known in grasshoppers and other orthopterans.

Stylets Needle-like form: applied to the various components of piercing mouthparts and also to a part of the sting of a bee or other hymenopteran.

Styli Small finger-like process.

Subcosta (Sc.) The second, usually unbranched, longitudinal wing vein, posterior to the costa.

Subgenital Below the genitalia.

Subimaginal Pertaining to the subimago, the first imaginal instar of mayflies.

Subimago First adult instar of the first imaginal instar of mayflies.

Subsist Manage to survive, survive, to keep alive

Sulcate Scored with usually longitudinal furrows.

Sulcus (pl. Sulci) Shallow furrows.

Supersonic Capable of exceeding speed of sound; capable of moving at a faster rate than sound

Suture A groove on the body surface which usually divides one plate or sclerite from the next: also the junction between the elytra of a beetle.

Swarms group of insects; a large group of insects, especially bees or gnats, in flight.

Symbiotic The state of living together of two dissimilar organisms in a mutually beneficial relationship.

Synthetically unnaturally, falsely, made artificially with chemicals.

Tannic acid Plant chemical used in tanning; acid made from the bark or galls of trees or plants and used in tanning leather and for making dyes, ink, and some medicines

Tarsal Relating to the group of bones forming the ankle joint tarsus ; having to do with the tarsus, the fifth segment of the insect's leg

Tarsi (pl. Tarsus) An insect's foot: primitively a single segment beyond the tibia, but usually consisting of several subdivisions or segments in most living insects.

Tarsomere Subdivision of the primary tarsal segment.

Tarsus (plural form of tarsi) Part of arthropod leg; fifth of six segments of an insect's leg and seventh of eight segments of a spider's leg; this segment bears the claws

Taxonomists Scientists who deal with the naming and classification plants, animals, and microorganisms.

Tegmen The leathery forewing of a grasshopper or similar insect, such as a cockroach

Tegmina (pl. Tegman) The leathery forewing of Orthopteroid insects..

Tegminization To become tegmina-like.

Telofilum The median tail filament, being a prolongation of tergum X1.

Tenaculum In Collembola, a minute appendage with two divergent prongs, situated medially on the ventral surface of the third abdominal segment, serving to hold the furcula in place.

Tentorium Endoskeleton of the head.

Terminal Of or relating to an end, extremity, or boundary.

Terrestrial Earthly, worldly, land-dwelling

Terrestrial Living on or in, or growing from land.

Test In most Coccids a scale covering, consisting of exuvial, secreted waxy material and excrement.

Thoracic Relating to the thorax

Thorax Upper part of animal's body. Part between head and abdomen.

Thorax The middle of the three major divisions of the insect body. The legs and wings (if present) are always attached to the thorax.

Tibia Inner bone of lower leg of animal; fifth of eight segments of a spider's leg; fourth of six segments of an insect's leg; in human beings, the tibia is the thicker of the two leg bones commonly referred to as the shinbone.

Tibia. The forth leg segment between the femur and the tarsus.

Tibiotarsus. In Collembola, the fourth segment of the leg.

Toxin-laden full of poison

Trachea (pl., Tracheae) One of the minute tubes which permeate the insect body and carry gases to and from the various organs etc. They open to the air at the spiracles.

Trachea an air tube in an insect; the windpipe in humans

Triangle A triangular region near the base of the dragonfly wing, often divided into smaller cells.

Trichobothrium A fine sensory seta.

Trilobite Extinct, marine arthropods; trilobites had jointed legs and a body divided into three vertical lobes and several horizontal segments

Triommatidium A reduced compound eye in apterous aphids with 3 separate lenses.

Triradiate Having three rays or radiating branches.

Trochanter Second of eight leg parts of a spider; second of six leg parts in insects

Trochanter The second segment of the leg, between coxa and femur: often very small and easily overlooked.

Trochantin Free sclerotized remnant located at the base of the leg, providing a second point of pleural articulation with the coxa.

Tropics A line of latitude on the Earth's globe either 23° 262 north of the equator tropic of Cancer or 23° 262 south tropic of Capricorn. Areas on the surface of the earth nearest the equator and farthest from the north and south polar regions

Truncate Ending abruptly: squared off.

Tsetse fly A two-winged biting fly that feeds on the blood of humans and animals and is responsible for transmitting several diseases, including sleeping sickness. Native to: central Africa. Genus: Glossina. Blood-sucking African flies that transmit sleeping sickness in humans

Tubercle A small knob-like or rounded protuberance.

Tuberculate Having tubercles: small knob-like or rounded protuberances.

Tuberculiform Having the form of a tubercle.

Turbinate Having a shaped like a top or an inverted cone.

Tylus The distal part of the clypeus in Hemiptera.

Tympanic Related to the eardrum or tympanum

Tympanum The auditory membrane or ear-drum of various insects.

Typhus Infectious diseases carried by fleas, lice, mites, and ticks that cause high fever, weakness, spots on the skin, and other symptoms

Ultraviolet light the invisible part of the light spectrum beyond the violet color

Vannal Same as anal area or lobe of wing.

Vegetation growing plants

Vein In insects, the rib-like tube that strengthen the wings.

Velcro A trademark for a fastener made from two strips

Venation The arrangement of veins in the wings of insects.

Venom Poison, toxin, antidote, poisonous matter normally secreted by some animals

Ventral Being or located near or on the lower surface.

Versatile Adaptable, resourceful, all-around, capable of doing many things well

Vertebrate Animal with backbone.

Vertex. The top of the head, between the eyes.

Vestigial Poorly developed, degenerate or atrophied, more fully functional in an earlier stage of development of the individual or species.

Vibrations Feelings, atmosphere, ambiance, rapid movements back and forth

Vinegaroon a common name for whipscorpions (not to be confused with true scorpions), an order of the Arachnida that have defensive fluids with the aroma of vinegar

Virus Bug, germ, organisms that produce disease; viruses depend on the living tissue of their hosts for their reproduction and growth

Voracious Insatiable, big, hungry, ravenous, gluttonous, greedy, rapacious, eating greedily

Warble To sing with trills and often changing notes; a bump under the hide on the back of an animal caused by the larva of a bot fly or warble fly

Wingpad. The undeveloped wing of nymphs and naiads, which appear as two flat structures on each side.

Wriggler Mosquito larva.

Yellow fever Fatal infectious disease; an infectious, often fatal viral disease of warm climates, transmitted by mosquitoes and marked by high fever, hemorrhaging, vomiting of blood, liver damage, and jaundice.

Yucca Flowering plant, An evergreen plant widely grown for its sharp lance-shaped leaves and clusters of white flowers that grow in vertical spikes. Native to: southwestern United States, Mexico. Genus: *Yucca*.

Bibliography

Arthur V. Evans and James N. Hogue, 2006, *Field Guide to Beetles of California*, University of California Press.

Barry, R.E., Jr., M.A. Botje, and L.B. Grantham. 1984. Vertical stratification of *Peromyscus leucopus* and *P. maniculatus* in southwestern Virginia. Journal of Mammalogy 65:145–148.

Bernhard Klausnitzer, *Insects: Their Biology and Cultural History*.

Borror, D.J. and DeLong, D.M. 1964. An Introduction to the Study of Insects. Holt, Rinehart and Winston.

Borror, D.J. and White, R.E. 1970. A Field Guide to Insects. Houghton Mifflin Co., Boston.

Boudreaux, H. B. 1979. Arthropod phylogeny, with special reference to insects. Wiley Interscience; New York, NY.

Carpenter, J. M., and W. C. Wheeler. 1999. Towards simultaneous analysis of morphological and molecular data in Hymenoptera. Zoologica Scripta 28(1-2):251-260.

Caudell, A.N. 1920. Zoraptera not an apterous order. Proceedings of the Entomological Society of Washington 22: 84-97.

Chao, R. F., and C. S. Chen. 2000. *Formosozoros newi*, a new genus and species of Zoraptera (Insecta) from Taiwan. Pan-Pacific Entomologist 76(1): 24-27.

Choe, J. C. 1989. *Zorotypus gurneyi*, new species, from Panama and redescription of *Zorotypus barberi* Gurney (Zoraptera,

Zorotypidae). Annals of the Entomological Society of America 82(2): 149-155.

Choe, J. C. 1994a. Sexual selection and mating system in *Zorotypus gurneyi* Choe (Insecta: Zoraptera). I. Dominance hierarchy and mating success. Behavioral Ecology and Sociobiology 34: 87-93.

Choe, J. C. 1994b. Sexual selection and mating system in *Zorotypus gurneyi* Choe (Insecta: Zoraptera). II. Determinants and dynamics of dominance. Behavioral Ecology and Sociobiology 34: 233-237.

Choe, J. C. 1995. Courtship feeding and repeated mating in *Zorotypus baberi* (Insecta: Zoraptera). Animal Behaviour 49: 1511-1520.

Choe, J. C. 1997. The evolution of mating systems in the Zoraptera: Mating variations and sexual conflicts. Pp. 130-145. *In* Choe, J. C., and B. J. Crespi (eds.), The evolution of mating systems in insects and arachnids. Cambridge University Press; Cambridge, UK.

Christopher O'Toole, *Firefly Encyclopedia of Insects and Spiders.*

Clausen, Lucy W. (1954). *Insect Fact and Folklore.* New York: Macmillan. XIV + 194 pp.

Craig, Owen (2001). "The Summer of Singing Cicadas". (February – Scribbly Gum – ABC Science Online). (accessed: December 23, 2006).

Craighead, F.C. 1950. Insect Enemies of Eastern Forests. U.S.D.A. Misc. Pub. No. 657.

CrossRef Catts, E.P. 1982. Biology of New World botflies: Cuterebridae. Annual Review of Entomology 27: 313–338.

CrossRef, CSA Burns, C.E., B.J. Goodwin, and R.S. Ostfeld. 2005. A prescription for longer life? Bot Fly parasitism of the White-footed Mouse. Ecology 86:753–761.

CrossRef, CSA Christopher, C.C., and G.W. Barrett. 2006. Coexistence of White-footed Mice (*Peromyscus leucopus*) and Golden Mice (*Ochrotomys nuttalli*) in a southestern forest. Journal of Mammalogy 87: 102–107.

Crozier, R. H. 1975. Hymenoptera. Borntraeger, Berlin; Stuttgart.

Danforth, B. D. 1989. The evolution of hymenopteran wings: The importance of size. Journal Of Zoology 218(2):247-276.

Egan, Rory B. (1994). *Cicada in Ancient Greece*. Third issue, November 1994.

Frost, S.W. 1959. Insect Life and Insect Natural History. Dover Publications Inc., New York.

Gauld, I. and Bolton, B. 1988. The Hymenoptera. British Museum (Natural History); Oxford University Press, London; Oxford; New York.

Gibson, G.A.P. 1985. Some pro- and mesothoracic structures important for phylogenetic analysis of Hymenoptera, with a review of terms used for the structures. Can. Ent. 117:1395-1443.

Goulet, H. and J. T. Huber. 1993. Hymenoptera of the world: an identification guide to families. Agriculture Canada Research Branch Monograph no. 1894E.

Hoppensteadt, Frank C. and Joseph B. Keller. (1976) *Synchronization of Periodical Cicada Emergences*. Science, Vol. 194, No. 4262 (Oct. 15, 1976), pp. 335–337.

Johnson, N. F. 1988. Midcoxal articulations and the phylogeny of the order Hymenoptera. Annals Of The Entomological Society Of America 81(6): 870-881.

Krombein, K. V. 1979. Catalog of Hymenoptera in America north of Mexico. Smithsonian Institution Press, Washington.

Myers, J.G. (1929). *Insect Singers: A Natural History of the Cicadas*. Routledge.

New, T. R. 2000. Zoraptera (Insecta) in East Malaysia: Notes on *Zorotypus caudelli* Karny. Oriental Insects 34: 77-82.

Poinar, G. O. 1988. *Zorotypus palaeus*, new species, a fossil Zoraptera (Insecta) in Dominican amber. Journal of the New York Entomological Society 96(3):253-259.

Quicke, D. L. J., M. G. Fitton, et al. 1992. Phylogenetic implications of the structure and distribution of ovipositor valvilli in the Hymenoptera (Insecta). Journal Of Natural History 26(3):587-608.

Quicke, D. L. J., S. N. Ingram, et al. 1992. Sperm structure and ultrastructure in the Hymenoptera (Insecta). Zoologica Scripta 21(4):381-402.

Ramel, Gordon (2005). *The Singing Cicadas.*

Rasnitsyn, A. P. 1998. On the taxonomic position of the insect order Zorotypida = Zoraptera. Zoologischer Anzeiger 237(2-3): 185-194.

Riegel, Garland (1994). *Cicada in Chinese Folklore.* Reproduced with permission from the Melsheimer Entomological Series. Third issue, November 1994.

Ronquist, F. 1999. Phylogeny of the Hymenoptera (Insecta): The state of the art. Zoologica Scripta 28(1-2):3-11.

Ronquist, F., A.P. Rasnitsyn, A. Roy, K. Eriksson, and M. Lindren. 1999. Phylogeny of the Hymenoptera: A cladistic reanalysis of Rasnitsyn's data (1988) data. Zoologica Scripta 28:13-50.

Saini, M. S. 1986. Comparative studies on the metapostnotal modifications in suborder Symphyta and Hymenoptera Parasitica. Journal Of Entomological Research 10(1):85-90.

Schulmeister, S. 2003. Morphology and evolution of the tarsal plantulae in Hymenoptera (Insecta), focussing on the basal lineages. Zoologica Scripta 32:153-172.

Schulmeister, S. 2003. Review of morphological evidence on the phylogeny of basal Hymenoptera (Insecta), with a discussion of the ordering of characters. Biological Journal of the Linnean Society 79:209-243.

Schulmeister, S. 2003. Simultaneous analysis of basal Hymenoptera (Insecta): introducing robust-choice sensitivity analysis. Biological Journal of the Linnean Society 79:245-275.

Smith EH, Whitman RC. 2003. *Ham/Cheese Skipper. In* National Pest Control Association Field Guide to Structural Pests: Stored Product Pests. Fairfax, VA.

Smith LF. 1943. Internal parasites of the red fox in Iowa. The Journal of Wildlife Management 7: 174-178.

Smith, R.C., Kelly, E.G., Dean, G.A., Bryson, H.R., and Parker, R.L. 1962. Insects in Kansas. Extension Division, Kansas State Univ., Manhattan, Kansas.

Triplehorn CA, Johnson NF. 2005. Borror and Delong's Introduction to the Study of Insects, 7th edition. Thomson Brooks/Cole Publishers. 864 pp.

Valentine, B. D. 1986. Grooming behavior in Embioptera and Zoraptera (Insecta). *Ohio Journal of Science 86(4): 150-152.*

Walker, Annette, *The Reed Handbook of Common New Zealand Insects*, Reed Books, 2000.

Wheeler, W. C., M. Whiting, Q. D. Wheeler, and J. M. Carpenter. 2001. The phylogeny of the extant hexapod orders. Cladistics 17: 113-169, 403-404.

William J. Bell, Louis M. Roth, and Christine A. Nalepa, *Cockroaches: Ecology, Behavior, and Natural History.*

Wilson, E. O. 1959. Some ecological characteristics of ants in New Guinea rain forests. Ecology 40: 437-447.

Zuzka J. 1978. Autogeny in *Piophila casei* (Diptera, Piophilidae). Folia Parasitologica, Prague 25: 173-77.

Index